CRITERIA FOR A RECOMMENDED STANDARD

Occupational Noise Exposure
Revised Criteria 1998

U.S. DEPARTMENT OF HEALTH AND HUMAN SERVICES
Public Health Service
Centers for Disease Control and Prevention
National Institute for Occupational Safety and Health
Cincinnati, Ohio

June 1998

DISCLAIMER

Mention of the name of any company or product does not constitute endorsement by the National Institute for Occupational Safety and Health.

Copies of this and other NIOSH documents are available from

Publications Dissemination
Education and Information Division
National Institute for Occupational Safety and Health
4676 Columbia Parkway
Cincinnati, OH 45226–1998

Fax number: (513) 533–8573
Telephone number: 1–800–35–NIOSH (1–800–356–4674)
E-mail: pubstaft@cdc.gov

To receive other information about occupational safety and health problems, call 1–800–35–NIOSH (1–800–356–4674), or visit the NIOSH Homepage on the World Wide Web at http://www.cdc.gov/niosh

DHHS (NIOSH) Publication No. 98-126

FOREWORD

In the Occupational Safety and Health Act of 1970 (Public Law 91-596), Congress declared that its purpose was to assure, so far as possible, safe and healthful working conditions for every working man and woman and to preserve our human resources. In this Act, the National Institute for Occupational Safety and Health (NIOSH) is charged with recommending occupational safety and health standards and describing exposure concentrations that are safe for various periods of employment—including but not limited to concentrations at which no worker will suffer diminished health, functional capacity, or life expectancy as a result of his or her work experience. By means of criteria documents, NIOSH communicates these recommended standards to regulatory agencies (including the Occupational Safety and Health Administration [OSHA]) and to others in the occupational safety and health community.

Criteria documents provide the scientific basis for new occupational safety and health standards. These documents generally contain a critical review of the scientific and technical information available on the prevalence of hazards, the existence of safety and health risks, and the adequacy of control methods. In addition to transmitting these documents to the Department of Labor, NIOSH also distributes them to health professionals in academic institutions, industry, organized labor, public interest groups, and other government agencies.

In 1972, NIOSH published *Criteria for a Recommended Standard: Occupational Exposure to Noise*, which provided the basis for a recommended standard to reduce the risk of developing permanent hearing loss as a result of occupational noise exposure [NIOSH 1972]. NIOSH has now evaluated the latest scientific information and has revised some of its previous recommendations. The 1998 recommendations go beyond attempting to conserve hearing by focusing on preventing occupational noise-induced hearing loss (NIHL).

The NIOSH recommended exposure limit (REL) for occupational noise exposure (85 decibels, A-weighted, as an 8-hour time-weighted average [85 dBA as an 8-hr TWA]) was reevaluated using contemporary risk assessment techniques and incorporating the 4000-hertz (Hz) audiometric frequency in the definition of hearing impairment. The new risk assessment reaffirms support for the 85-dBA REL. With a 40-year lifetime exposure at the 85-dBA REL, the excess risk of developing occupational NIHL is 8%—considerably lower than the 25% excess risk at the 90-dBA permissible exposure limit (PEL) currently enforced by the Occupational Safety and Health Administration (OSHA) and the Mine Safety and Health Administration (MSHA).

NIOSH previously recommended an exchange rate of 5 dB for the calculation of time-weighted average (TWA) exposures to noise. However, NIOSH now recommends a 3-dB exchange rate, which is more firmly supported by scientific evidence. The 5-dB exchange rate is still used by OSHA and MSHA, but the 3-dB exchange rate has been increasingly supported by national and international consensus.

NIOSH recommends an improved criterion for significant threshold shift: an increase of 15 dB in the hearing threshold level (HTL) at 500, 1000, 2000, 3000, 4000, or 6000 Hz in either ear, as determined by two consecutive audiometric tests. The new criterion has the advantages of a high identification rate and a low false-positive rate. In comparison, the criterion NIOSH recommended in 1972 has a high false-positive rate, and the OSHA criterion (called the standard threshold shift) has a relatively low identification rate.

In contrast with the 1972 criterion, the new NIOSH criterion no longer recommends age correction on individual audiograms. This practice is not scientifically valid and would delay intervention to prevent further hearing losses in workers whose HTLs have increased because of occupational noise exposure. OSHA currently allows age correction only as an option.

The noise reduction rating (NRR) is a single-number, laboratory-derived rating that the U.S. Environmental Protection Agency (EPA) requires to be shown on the label of each hearing protector sold in the United States. In calculating the noise exposure to the wearer of a hearing protector at work, OSHA derates the NRR by one-half for all types of hearing protectors. In 1972, NIOSH recommended the use of the full NRR value; however, in this document, NIOSH recommends derating by subtracting from the NRR 25%, 50%, and 70% for earmuffs, formable earplugs, and all other earplugs, respectively. This variable derating scheme, as opposed to OSHA's straight derating scheme, considers the performances of different types of hearing protectors.

This document also provides recommendations for the management of hearing loss prevention programs (HLPPs) for workers whose noise exposures equal or exceed 85 dBA. The recommendations include program evaluation, which was not articulated in the 1972 criteria document and is not included in the OSHA and MSHA standards.

Adherence to the revised recommended noise standard will minimize the risk of developing occupational NIHL.

Linda Rosenstock, M.D., M.P.H.
Director, National Institute for
  Occupational Safety and Health
Centers for Disease Control and Prevention

ABSTRACT

This criteria document reevaluates and reaffirms the recommended exposure limit (REL) for occupational noise exposure established by the National Institute for Occupational Safety and Health (NIOSH) in 1972. The REL is 85 decibels, A-weighted, as an 8-hr time-weighted average (85 dBA as an 8-hr TWA). Exposures at or above this level are hazardous.

By incorporating the 4000-Hz audiometric frequency into the definition of hearing impairment in the risk assessment, NIOSH has found an 8% excess risk of developing occupational noise-induced hearing loss (NIHL) during a 40-year lifetime exposure at the 85-dBA REL. NIOSH has also found that scientific evidence supports the use of a 3-dB exchange rate for the calculation of TWA exposures to noise.

The recommendations in this document go beyond attempts to conserve hearing by focusing on prevention of occupational NIHL. For workers whose noise exposures equal or exceed 85 dBA, NIOSH recommends a hearing loss prevention program (HLPP) that includes exposure assessment, engineering and administrative controls, proper use of hearing protectors, audiometric evaluation, education and motivation, recordkeeping, and program audits and evaluations.

Audiometric evaluation is an important component of an HLPP. To provide early identification of workers with increasing hearing loss, NIOSH has revised the criterion for significant threshold shift to an increase of 15 dB in the hearing threshold level (HTL) at 500, 1000, 2000, 3000, 4000, or 6000 Hz in either ear, as determined by two consecutive tests. To permit timely intervention and prevent further hearing losses in workers whose HTLs have increased because of occupational noise exposure, NIOSH no longer recommends age correction on individual audiograms.

CONTENTS

Foreword iii
Abstract v
Abbreviations x
Glossary xii
Acknowledgments xvi

1 Recommendations for a Noise Standard 1

1.1 Recommended Exposure Limit (REL) 1
1.1.1 Exposure Levels and Durations 1
1.1.2 Time-Weighted Average (TWA) 1
1.1.3 Daily Noise Dose 1
1.1.4 Ceiling Limit 4

1.2 Hearing Loss Prevention Program 4

1.3 Noise Exposure Assessment 4
1.3.1 Initial Monitoring 4
1.3.2 Periodic Monitoring 4
1.3.3 Instrumentation 4

1.4 Engineering and Administrative Controls and Work Practices 5

1.5 Hearing Protectors 5

1.6 Medical Surveillance 5
1.6.1 Audiometry 6
1.6.2 Baseline Audiogram 6
1.6.3 Monitoring Audiogram and Retest Audiogram 7
1.6.4 Confirmation Audiogram, Significant Threshold Shift,
and Follow-up Action 7
1.6.5 Exit Audiogram 8

1.7 Hazard Communication 8
1.7.1 Warning Signs 8
1.7.2 Notification to Workers 8

1.8 Training 8

1.9 Program Evaluation Criteria 9

1.10 Recordkeeping 9
1.10.1 Exposure Assessment Records 9
1.10.2 Medical Surveillance Records 10

1.10.3 Record Retention... 10
1.10.4 Availability of Records... 10
1.10.5 Transfer of Records... 10

1.11 ANSI Standards... 10

2 Introduction... 11

2.1 Recognition of Noise as a Health Hazard... 11

2.2 Noise-Induced Hearing Loss (NIHL)... 11

2.3 Physical Properties of Sound... 12

2.4 Number of Noise-Exposed Workers in the United States... 12

2.5 Legislative History... 13

2.6 Scope of This Revision of the Noise Criteria Document... 18

3 Basis for the Exposure Standard... 19

3.1 Quantitative Risk Assessment... 19

3.1.1 NIOSH Risk Assessment in 1972... 19
3.1.2 NIOSH Risk Assessment in 1997... 20

3.2 Ceiling Limit... 24

3.3 Exchange Rate... 25

3.4 Impulsive Noise... 29

3.4.1 Evidence That Impulsive Noise Effects Do Not Conform to the Equal-Energy Rule... 30
3.4.2 Evidence That Impulsive Noise Effects Conform to the Equal-Energy Rule... 31
3.4.3 Combined Exposure to Impulsive and Continuous-Type Noises... 32

4 Instrumentation for Noise Measurement... 33

4.1 Sound Level Meter... 33

4.1.1 Frequency Weighting Networks... 33
4.1.2 Exponential Time Weighting... 33
4.1.3 Microphones for Sound Level Meters... 34

4.2 Noise Dosimeter... 35

4.3 Range of Sound Levels... 35

5 Hearing Loss Prevention Programs (HLPPs)... 36

5.1 Personnel Requirements... 37

5.2 Initial and Annual Audits (Component 1)... 38

5.3 Exposure Assessment (Component 2)... 38

5.4 Engineering and Administrative Controls (Component 3) 40
5.5 Audiometric Evaluation and Monitoring (Component 4) 41
    5.5.1 Audiometry 41
        5.5.1.1 Baseline Audiogram 49
        5.5.1.2 Monitoring Audiograms 49
        5.5.1.3 Retest Audiograms 49
        5.5.1.4 Confirmation Audiograms 50
        5.5.1.5 Exit Audiogram 50
    5.5.2 Audiometers 50
5.6 Use of Hearing Protectors (Component 5) 52
5.7 Education and Motivation (Component 6) 52
5.8 Recordkeeping (Component 7) 55
    5.8.1 Noise Exposure Records 55
    5.8.2 Audiometric Records 56
    5.8.3 Hearing Protection Records 56
    5.8.4 Education Records 57
    5.8.5 Other Records 57
5.9 Evaluation of Program Effectiveness (Component 8) 57
    5.9.1 Individual Effectiveness 57
    5.9.2 Overall Program Effectiveness 57
5.10 Age Correction 59

6 Hearing Protectors 61

7 Research Needs 69
7.1 Noise Control 69
7.2 Impulsive Noise 69
7.3 Nonauditory Effects 69
7.4 Auditory Effects of Ototoxic Chemical Exposures 70
7.5 Exposure Monitoring 70
7.6 Hearing Protectors 70
7.7 Training and Motivation 71
7.8 Program Evaluation 71
7.9 Rehabilitation 71

References 73

Appendix 91

ABBREVIATIONS

AAO-HNS	American Academy of Otolaryngology-Head and Neck Surgery
AIHA	American Industrial Hygiene Association
ANSI	American National Standards Institute
AOMA	American Occupational Medical Association
ASHA	American Speech-Language-Hearing Association
CAOHC	Council for Accreditation in Occupational Hearing Conservation
CFR	*Code of Federal Regulations*
CHABA	Committee on Hearing, Bioacoustics, and Biomechanics
CI	confidence interval
dB	decibel(s)
dB SPL	decibel(s), sound pressure level
dBA	decibel(s), A-weighted
EPA	U.S. Environmental Protection Agency
Fed. Reg.	*Federal Register*
HLPP	hearing loss prevention program
hr	hour(s)
HTL	hearing threshold level
Hz	hertz
ISO	International Standards Organization
kHz	kilohertz
$L_{Aeq\,8\,hr}$	equivalent continuous sound for 8 hr
min	minute(s)

Abbreviations

ms	millisecond(s)
MSHA	Mine Safety and Health Administration
NHANES	National Health and Nutrition Examination Survey
NHCA	National Hearing Conservation Association
NIHL	noise-induced hearing loss
NIOSH	National Institute for Occupational Safety and Health
NOES	National Occupational Exposure Survey
NOHSM	National Occupational Health Survey of Mining
NRR	noise reduction rating
ONHS	Occupational Noise and Hearing Survey
OSHA	Occupational Safety and Health Administration
PEL	permissible exposure limit
REAT	real ear attenuation at threshold
REL	recommended exposure limit
s	second(s)
SIC	standard industrial classification
SPL	sound pressure level
STS	standard threshold shift
T-BEAM	task-based exposure assessment model
TTS_2	temporary threshold shift 2 min after a period of noise exposure
TWA	time-weighted average

GLOSSARY

Where possible, the definition is quoted from the appropriate American National Standards Institute (ANSI) standard, ANSI S1.1-1994 [ANSI 1994] or ANSI S3.20-1995 [ANSI 1995], under the term(s) used in that standard.

Audiogram: Graph of hearing threshold level as a function of frequency (ANSI S3.20-1995: audiogram).

Baseline audiogram: The audiogram obtained from an audiometric examination administered before employment or within the first 30 days of employment that is preceded by a period of at least 12 hr of quiet. The baseline audiogram is the audiogram against which subsequent audiograms will be compared for the calculation of significant threshold shift.

Continuous noise: Noise with negligibly small fluctuations of level within the period of observation (ANSI S3.20-1995: stationary noise; steady noise).

Crest factor: Ten times the logarithm to the base ten of the square of the wideband peak amplitude of a signal to the time-mean-square amplitude over a stated time period. Unit, dB (ANSI S3.20-1995: crest factor).

Decibel (dB): Unit of level when the base of the logarithm is the 10th root of 10 and the quantities concerned are proportional to power (ANSI S1.1-1994: decibel).

Decibel, A-weighted (dBA): Unit representing the sound level measured with the A-weighting network on a sound level meter. (Refer to Table 4-1 for the characteristics of the weighting networks.)

Decibel, C-weighted (dBC): Unit representing the sound level measured with the C-weighting network on a sound level meter. (Refer to Table 4-1 for the characteristics of the weighting networks.)

Derate: To use a fraction of a hearing protector's noise reduction rating (NRR) to calculate the noise exposure of a worker wearing that hearing protector. (See NRR below.)

Dose: The amount of actual exposure relative to the amount of allowable exposure, and for which 100% and above represents exposures that are hazardous. The noise dose is calculated according to the following formula:

$$D = [C_1/T_1 + C_2/T_2 + \ldots + C_n/T_n] \times 100$$

Where

C_n = total time of exposure at a specified noise level
T_n = exposure time at which noise for this level becomes hazardous

Effective noise level: The estimated A-weighted noise level at the ear when wearing hearing protectors. Effective noise level is computed by (1) subtracting derated NRRs from C-weighted noise exposure levels, or (2) subtracting derated NRRs minus 7 dB from A-weighted noise exposure levels. Unit, dB. (See Appendix.)

Equal-energy hypothesis: A hypothesis stating that equal amounts of sound energy will produce equal amounts of hearing impairment, regardless of how the sound energy is distributed in time.

Equivalent continuous sound level: Ten times the logarithm to the base ten of the ratio of time-mean-square instantaneous A-weighted sound pressure, during a stated time interval T, to the square of the standard reference sound pressure. Unit, dB; respective abbreviations, TAV and TEQ; respective letter symbols, L_{AT} and L_{AeqT} (ANSI S1.1-1994: time-average sound level; time-interval equivalent continuous sound level; time-interval equivalent continuous A-weighted sound pressure level; equivalent continuous sound level).

Excess risk: Percentage with material impairment of hearing in an occupational-noise-exposed population after subtracting the percentage who would normally incur such impairment from other causes in a population not exposed to occupational noise.

Exchange rate: An increment of decibels that requires the halving of exposure time, or a decrement of decibels that requires the doubling of exposure time. For example, a 3-dB exchange rate requires that noise exposure time be halved for each 3-dB increase in noise level; likewise, a 5-dB exchange rate requires that exposure time be halved for each 5-dB increase.

Fence: The hearing threshold level above which a material impairment of hearing is considered to have occurred.

Frequency: For a function periodic in time, the reciprocal of the period. Unit, hertz (Hz) (ANSI S1.1-1994: frequency).

Hearing threshold level (HTL): For a specified signal, amount in decibels by which the hearing threshold for a listener, for one or both ears, exceeds a specified reference equivalent threshold level. Unit, dB (ANSI S1.1-1994: hearing level; hearing threshold level).

Immission level: A descriptor for noise exposure, in decibels, representing the total sound energy incident on the ear over a specified period of time (e.g., months, years).

Impact: Single collision of one mass in motion with a second mass that may be in motion or at rest (ANSI S1.1-1994: impact).

Impulse: Product of a force and the time during which the force is applied; more specifically, impulse is the time integral of force from an initial time to a final time, the force being time-dependent and equal to zero before the initial time and after the final time (ANSI S1.1–1994: impulse).

Impulsive noise: Impulsive noise is characterized by a sharp rise and rapid decay in sound levels and is less than 1 sec in duration. For the purposes of this document, it refers to impact or impulse noise.

Intermittent noise: Noise levels that are interrupted by intervals of relatively low sound levels.

Noise: (1) Undesired sound. By extension, noise is any unwarranted disturbance within a useful frequency band, such as undesired electric waves in a transmission channel or device. (2) Erratic, intermittent, or statistically random oscillation (ANSI S1.1–1994: noise).

Noise reduction rating (NRR): The NRR, which indicates a hearing protector's noise reduction capabilities, is a single-number rating that is required by law to be shown on the label of each hearing protector sold in the United States. Unit, dB.

Permanent threshold shift (PTS): Permanent increase in the threshold of audibility for an ear. Unit, dB (ANSI S3.20–1995: permanent threshold shift; permanent hearing loss; PTS).

Pulse range: Difference in decibels between the peak level of an impulsive signal and the root-mean-square level of a continuous noise.

Significant threshold shift: A shift in hearing threshold, outside the range of audiometric testing variability (±5 dB), that warrants followup action to prevent further hearing loss. NIOSH defines significant threshold shift as an increase in the HTL of 15 dB or more at any frequency (500, 1000, 2000, 3000, 4000, or 6000 Hz) in either ear that is confirmed for the same ear and frequency by a second test within 30 days of the first test.

Sound: (1) Oscillation in pressure, stress, particle displacement, particle velocity, etc. in a medium with internal forces (e.g., elastic or viscous), or the superposition of such propagated oscillations. (2) Auditory sensation evoked by the oscillation described above (ANSI S1.1–1994: sound).

Sound intensity: Average rate of sound energy transmitted in a specified direction at a point through a unit area normal to this direction at the point considered. Unit, watt per square meter (W/m^2); symbol, I (ANSI S1.1–1994: sound intensity; sound-energy flux density; sound-power density).

Sound intensity level: Ten times the logarithm to the base ten of the ratio of the intensity of a given sound in a stated direction to the reference sound intensity of 1 picoWatt per square meter (pW/m^2). Unit, dB; symbol, L (ANSI S1.1–1994: sound intensity level).

Sound pressure: Root-mean-square instantaneous sound pressure at a point during a given time interval. Unit, Pascal (Pa) (ANSI S1.1–1994: sound pressure; effective sound pressure).

Sound pressure level: (1) Ten times the logarithm to the base ten of the ratio of the time-mean-square pressure of a sound, in a stated frequency band, to the square of the reference sound pressure in gases of 20 micropascals (μPa). Unit, dB; symbol, L_p. (2) For sound in media other than gases, unless otherwise specified, reference sound pressure in 1 μPa (ANSI S1.1–1994: sound pressure level).

Temporary threshold shift: Temporary increase in the threshold of audibility for an ear caused by exposure to high-intensity acoustic stimuli. Such a shift may be caused by other means such as use of aspirin or other drugs. Unit, dB. (ANSI S3.20–1995: temporary threshold shift; temporary hearing loss).

Time-weighted average (TWA): The averaging of different exposure levels during an exposure period. For noise, given an 85-dBA exposure limit and a 3-dB exchange rate, the TWA is calculated according to the following formula:

$$TWA = 10.0 \times Log(D/100) + 85$$

where D = dose.

Varying noise: Noise, with or without audible tones, for which the level varies substantially during the period of observation (ANSI S3.20–1995: nonstationary noise; nonsteady noise; time-varying noise).

ACKNOWLEDGMENTS

This document was prepared by the staff of the National Institute for Occupational Safety and Health (NIOSH). Principal responsibility for this document rested with the Education and Information Division, Paul A. Schulte, Ph.D., Director, and the Division of Biomedical and Behavioral Science, Derek E. Dunn, Ph.D., Director. Henry S. Chan was the document manager. John R. Franks, Ph.D., Carol J. Merry, Ph.D., Mark R. Stephenson, Ph.D., and Christa L. Themann contributed the principal input on the technical aspects of noise measurements, noise health effects, and the requisite components of a hearing loss prevention program. Mary M. Prince, Ph.D., Randall J. Smith, Leslie T. Stayner, Ph.D., and Stephen J. Gilbert provided risk assessment and statistical calculations. Barry Lempert recovered and reformatted the Occupational Noise and Hearing Survey (ONHS) data. David H. Pedersen, Ph.D. and Randy O. Young provided data from the National Occupational Exposure Survey (NOES). Dennis W. Groce and Janet M. Hale provided data from the National Occupational Health Survey of Mining (NOHSM). Ralph D. Zumwalde and Marie Haring Sweeney, Ph.D., provided policy review. Robert J. Tuchman, Anne C. Hamilton, Jane Weber, and Susan Feldmann edited the document. Susan Kaelin and Vanessa Becks provided editorial assistance and desktop publishing. Judy C. Curless, Sharon L. Cheesman, and Michelle Brunswick provided word processing and production support.

NIOSH gratefully acknowledges the contributions of Alice H. Suter, Ph.D. (Alice Suter and Associates, Ashland, OR) and Julia D. Royster, Ph.D. (Environmental Noise Consultants, Inc., Raleigh, NC), who served as consultants in the areas of the 3-decibel exchange rate and criteria for significant threshold shift, respectively.

NIOSH thanks the following consultants for their participation in the public meeting held on June 20–21, 1997, in Cincinnati, OH:

Henning von Gierke, Dr. Eng.
Yellow Springs, OH

Daniel L. Johnson, Ph.D
Interactive Acoustics
Provo, UT

Scott Schneider
Center to Protect Workers' Rights
Washington, DC

Anne R. Shields, Ph.D.
Aberdeen Proving Ground, MD

Thomas Simpson, Ph.D.
Wayne State University
Detroit, MI

Alice H. Suter, Ph.D.
Alice Suter and Associates
Ashland, OR

Edwin Toothman
Noise/Hearing Construction
Bethlehem, PA

We also thank James E. Lankford, Ph.D. (Northern Illinois University, DeKalb, IL) and Charles W. Nixon, Ph.D. (Wright Patterson Air Force Base, Dayton, OH) for reviewing the draft.

CHAPTER 1

Recommendations for a Noise Standard

The National Institute for Occupational Safety and Health (NIOSH) recommends the following standard for promulgation by regulatory agencies such as the Occupational Safety and Health Administration (OSHA) and the Mine Safety and Health Administration (MSHA) to protect workers from hearing losses resulting from occupational noise exposure. If this recommended standard is promulgated by a regulatory agency, the mandatory and nonmandatory provisions of the standard are indicated by the words *shall* and *should*, respectively.

1.1 Recommended Exposure Limit (REL)

The NIOSH recommended exposure limit (REL) for occupational noise exposure encompasses the provisions in Sections 1.1.1 through 1.1.4. The REL is 85 decibels, A-weighted, as an 8-hr time-weighted average (85 dBA as an 8-hr TWA). Exposures at and above this level are considered hazardous.

1.1.1 Exposure Levels and Durations

Occupational noise exposure shall be controlled so that worker exposures are less than the combination of exposure level (L) and duration (T), as calculated by the following formula (or as shown in Table 1-1).

$$T(\min) = \frac{480}{2^{(L-85)/3}}$$

where 3 = the exchange rate.

1.1.2 Time-Weighted Average (TWA)

In accordance with Section 1.1.1, the REL for an 8-hr work shift is a TWA of 85 dBA using a 3-decibel (dB) exchange rate.

1.1.3 Daily Noise Dose

When the daily noise exposure consists of periods of different noise levels, the daily dose (D) shall not equal or exceed 100, as calculated according to the following formula:

$$D = [C_1/T_1 + C_2/T_2 + ... + C_n/T_n] \times 100$$

where

C_n = total time of exposure at a specified noise level, and
T_n = exposure duration for which noise at this level becomes hazardous.

The daily dose can be converted into an 8-hr TWA according to the following formula (or as shown in Table 1–2):

$$TWA = 10.0 \times \mathrm{Log}(D/100) + 85$$

Table 1–1. Combinations of noise exposure levels and durations that no worker exposure shall equal or exceed

Exposure level, L (dBA)	Duration, T			Exposure level, L (dBA)	Duration, T		
	Hours	Minutes	Seconds		Hours	Minutes	Seconds
80	25	24	—	106	—	3	45
81	20	10	—	107	—	2	59
82	16	—	—	108	—	2	22
83	12	42	—	109	—	1	53
84	10	5	—	110	—	1	29
85	8	—	—	111	—	1	11
86	6	21	—	112	—	—	56
87	5	2	—	113	—	—	45
88	4	—	—	114	—	—	35
89	3	10	—	115	—	—	28
90	2	31	—	116	—	—	22
91	2	—	—	117	—	—	18
92	1	35	—	118	—	—	14
93	1	16	—	119	—	—	11
94	1	—	—	120	—	—	9
95	—	47	37	121	—	—	7
96	—	37	48	122	—	—	6
97	—	30	—	123	—	—	4
98	—	23	49	124	—	—	3
99	—	18	59	125	—	—	3
100	—	15	—	126	—	—	2
101	—	11	54	127	—	—	1
102	—	9	27	128	—	—	1
103	—	7	30	129	—	—	1
104	—	5	57	130–140	—	—	<1
105	—	4	43	—	—	—	—

Table 1–2. Daily noise dose as an 8-hr TWA*

Dose (%)	dBA as 8-hr TWA	Dose (%)	dBA as 8-hr TWA	Dose (%)	dBA as 8-hr TWA
20	78.0	2,000	98.0	450,000	121.5
30	79.8	2,500	99.0	500,000	122.0
40	81.0	3,000	99.8	600,000	122.8
50	82.0	3,500	100.4	700,000	123.5
60	82.8	4,000	101.0	800,000	124.0
70	83.5	4,500	101.5	900,000	124.5
80	84.0	5,000	102.0	1,000,000	125.0
90	84.5	6,000	102.8	1,100,000	125.4
100	85.0	7,000	103.5	1,200,000	125.8
110	85.4	8,000	104.0	1,300,000	126.1
120	85.8	9,000	104.5	1,400,000	126.5
130	86.1	10,000	105.0	1,600,000	127.0
140	86.5	12,000	105.8	1,800,000	127.6
150	86.8	14,000	106.5	2,000,000	128.0
170	87.3	16,000	107.0	2,200,000	128.4
200	88.0	18,000	107.6	2,400,000	128.8
250	89.0	20,000	108.0	2,600,000	129.1
300	89.8	25,000	109.0	2,800,000	129.5
350	90.4	30,000	109.8	3,000,000	129.8
400	91.0	35,000	110.4	3,500,000	130.4
450	91.5	40,000	111.0	4,000,000	131.0
500	92.0	45,000	111.5	4,500,000	131.5
550	92.4	50,000	102.0	5,000,000	132.0
600	92.8	60,000	112.8	6,000,000	132.8
650	93.1	70,000	113.5	7,000,000	133.5
700	93.5	80,000	114.0	8,000,000	134.0
750	93.8	90,000	114.5	9,000,000	134.5
800	94.0	100,000	115.0	10,000,000	135.0
900	94.5	110,000	115.4	12,000,000	135.8
1,000	95.0	120,000	115.8	14,000,000	136.5
1,050	95.2	130,000	116.1	16,000,000	137.0
1,100	95.4	140,000	116.5	18,000,000	137.6
1,150	95.6	150,000	116.8	20,000,000	138.0
1,200	95.8	175,000	117.4	22,000,000	138.4
1,300	96.1	200,000	118.0	24,000,000	138.8
1,400	96.5	225,000	118.5	26,000,000	139.0
1,500	96.8	250,000	119.0	28,000,000	139.5
1,600	97.0	275,000	119.4	30,000,000	139.8
1,700	97.3	300,000	119.8	32,500,000	140.1
1,800	97.6	350,000	120.4		
1,900	97.8	400,000	121.0		

*TWA = $10 \times \text{Log}(D/100) + 85$

1.1.4 Ceiling Limit

Exposure to continuous, varying, intermittent, or impulsive noise shall not exceed 140 dBA.

1.2 Hearing Loss Prevention Program

The employer shall institute an effective hearing loss prevention program (HLPP) described in Sections 1.3 through 1.11 when any worker's 8-hr TWA exposure equals or exceeds 85 dBA.

1.3 Noise Exposure Assessment

The employer shall conduct a noise exposure assessment when any worker's 8-hr TWA exposure equals or exceeds 85 dBA. Exposure measurements shall conform to the *American National Standard Measurement of Occupational Noise Exposure*, ANSI S12.19–1996 [ANSI 1996a]. Noise exposure is to be measured without regard for the wearing of hearing protectors.

1.3.1 Initial Monitoring

When a new HLPP is initiated, an initial monitoring of the worksite or of noisy work tasks shall be conducted to determine the noise exposure levels representative of all workers whose 8-hr TWA noise exposures may equal or exceed 85 dBA. For workers remaining in essentially stationary, continuous noise levels, either a sound level meter or a dosimeter may be used. However, for workers who move around frequently or who perform different tasks with intermittent or varying noise levels, a task-based exposure monitoring strategy may provide a more accurate assessment of the extent of exposures.

1.3.2 Periodic Monitoring

If any worker's 8-hr TWA exposure to noise equals or exceeds 85 dBA, monitoring shall be repeated at least every 2 years. Monitoring shall be repeated within 3 months of the occurrence when there is a change in equipment, production processes or maintenance routines. It may also be prudent to assess noise exposures when work practices have changed and/or if workers are developing significant threshold shifts (see Section 1.6.4).

1.3.3 Instrumentation

Instruments used to measure workers' noise exposures shall be calibrated to ensure measurement accuracy and, at a minimum, they shall conform to the *American National Standard Specification for Sound Level Meters*, ANSI S1.4–1983 and S1.4A–1985, Type 2 [ANSI 1983, 1985] or, with the exception of the operating range, to the *American National Standard Specification for Personal Noise Dosimeters*, ANSI S1.25–1991 [ANSI 1991a]. If a sound level meter is used, the meter response shall be set at SLOW.

In determining TWA exposures, all continuous, varying, intermittent, and impulsive sound levels from 80 to 140 dBA shall be integrated into the noise measurements.

1.4 Engineering and Administrative Controls and Work Practices

To the extent feasible, engineering controls, administrative controls, and work practices shall be used to ensure that workers are not exposed to noise at or above 85 dBA as an 8-hr TWA. The use of administrative controls shall not result in exposing more workers to noise.

1.5 Hearing Protectors

Workers shall be required to wear hearing protectors when engaged in work that exposes them to noise that equals or exceeds 85 dBA as an 8-hr TWA.[*] The employer shall provide hearing protectors at no cost to the workers.

Hearing protectors shall attenuate noise sufficiently to keep the worker's "real-world" exposure (i.e., the noise exposure at the worker's ear when hearing protectors are worn) below 85 dBA as an 8-hr TWA. Workers whose 8-hr TWA exposures exceed 100 dBA should wear double hearing protection (i.e., they should wear earplugs and earmuffs simultaneously).[†]

To compensate for known differences between laboratory-derived attenuation values and the protection obtained by a worker in the real world, the labeled noise reduction ratings (NRRs) shall be derated as follows: (1) earmuffs—subtract 25% from the manufacturers' labeled NRR; (2) slow-recovery formable earplugs—subtract 50%; and (3) all other earplugs—subtract 70% from the manufacturers' labeled NRR. These derating values shall be used until such time as manufacturers test and label their products in accordance with a subject-fit method such as method B of ANSI S12.6–1997, *American National Standard Methods for Measuring the Real-Ear Attenuation of Hearing Protectors* [ANSI 1997]. Chapter 6 (p. 62) describes methods for using the NRR.

The employer shall train workers at least annually to select, fit, and use a variety of appropriate hearing protectors. By making a variety of devices available and training the workers in their use, the employer will substantially increase the likelihood that hearing protector use will be effective and worthwhile.

1.6 Medical Surveillance

The employer shall provide audiometry for all workers whose exposures equal or exceed 85 dBA as an 8-hr TWA.

---

[*]This recommendation should not be construed to imply that workers need not wear hearing protection unless their 8-hr TWAs equal or exceed 85 dBA. For example, it would be prudent for a worker in and out of noise or habitually exposed to loud noise (e.g., 91 dBA for 1 hr and 59 min) to wear hearing protection while in noise—even though his or her dose was less than 100%.

[†]The intent of this section is not to advocate hearing protectors as the primary means of control; however, when engineering controls, administrative controls, and work practices cannot keep workers' exposures below 85 dBA as an 8-hr TWA, the use of hearing protectors shall be required. For most TWA exposures exceeding 105 dBA, hearing protectors will be necessary to supplement engineering and administrative controls.

Noise Exposure

1.6.1 Audiometry

Audiometric tests shall be performed by a physician, an audiologist, or an occupational hearing conservationist certified by the Council for Accreditation in Occupational Hearing Conservation (CAOHC) or the equivalent, working under the supervision of an audiologist or physician. The appropriate professional notation (e.g., licensure, certification, or CAOHC certification number) shall be recorded on each worker's audiogram.

Audiometric testing shall consist of air-conduction, pure-tone, hearing threshold measures at no less than 500, 1000, 2000, 3000, 4000, and 6000 hertz (Hz). Right and left ears shall be individually tested. The 8000-Hz threshold should also be tested as an option and as a useful source of information about the etiology of a hearing loss.

Audiometric tests shall be conducted with audiometers that meet the specifications of and are maintained and used in accordance with the *American National Standard Specifications for Audiometers*, ANSI S3.6-1996 [ANSI 1996b]. Audiometers shall receive a daily functional check, an acoustic calibration check whenever the functional check indicates a threshold difference exceeding 10 dB in either earphone at any frequency, and an exhaustive calibration check annually or whenever an acoustic calibration indicates the need—as outlined in Section 5.5.2. The date of the last annual calibration shall be recorded on each worker's audiogram.

Audiometric tests shall be conducted in a room where ambient noise levels conform to all requirements of the *American National Standard Maximum Permissible Ambient Noise Levels for Audiometric Test Rooms*, ANSI S3.1-1991 [ANSI 1991b]. Instruments used to measure ambient noise shall conform to the *American National Standard Specification for Sound Level Meters*, ANSI S1.4-1983 and S1.4A-1985, Type 1 [ANSI 1983, 1985] and the *American National Standard Specification for Octave-Band and Fractional-Octave-Band Analog and Digital Filters*, ANSI S1.11-1986 [ANSI 1986]. For permanent onsite testing facilities, ambient noise levels shall be checked at least annually. For mobile testing facilities, ambient noise levels shall be tested daily or each time the facility is moved, whichever is more often. Ambient noise measurements shall be obtained under conditions representing the typical acoustical environment likely to be present when audiometric testing is performed. Ambient noise levels shall be recorded on each audiogram or made otherwise accessible to the professional reviewer of the audiograms.

1.6.2 Baseline Audiogram

A baseline audiogram shall be obtained before employment or within 30 days of employment for all workers who must be enrolled in the HLPP. Workers shall not be exposed to noise levels at or above 85 dBA for a minimum of 12 hr before receiving a baseline audiometric test. Hearing protectors shall not be used in lieu of the required quiet period.

1.6.3 Monitoring Audiogram and Retest Audiogram

All workers enrolled in the HLPP shall have their hearing threshold levels (HTLs) measured annually. These audiometric tests shall be conducted during the worker's normal work shift. This audiogram shall be referred to as the "monitoring audiogram." The monitoring audiogram shall be examined immediately to determine whether a worker has a change in hearing relative to his or her baseline audiogram.

When the monitoring audiogram detects a change in the HTL in either ear that equals or exceeds 15 dB at 500, 1000, 2000, 3000, 4000, or 6000 Hz, an optional retest may be conducted immediately to determine whether the significant threshold shift is persistent. In most cases, the retest will demonstrate that the worker does *not* have a persistent threshold shift, thereby eliminating the need for a confirmation audiogram and followup action. If a persistent threshold shift *has* occurred, the worker shall be informed that his or her hearing may have worsened and additional hearing tests will be necessary.

1.6.4 Confirmation Audiogram, Significant Threshold Shift, and Followup Action

When a worker's monitoring audiogram detects a threshold shift as outlined in Section 1.6.3, he or she shall receive a confirmation audiogram within 30 days. This confirmation test shall be conducted under the same conditions as those of a baseline audiometric test. If the confirmation audiogram shows the persistence of a threshold shift, the audiograms and other appropriate records shall be reviewed by an audiologist or physician.

If this review validates the threshold shift, the threshold shift is considered to be a significant threshold shift. This shift shall be recorded in the worker's medical record, and the confirmation audiogram shall serve as the new baseline and shall be used to calculate any subsequent significant threshold shift. Whenever possible, the worker should receive immediate feedback on the results of his or her hearing test; however, in no case shall the worker be required to wait more than 30 days.

When a significant threshold shift has been validated, the employer shall take appropriate action to protect the worker from additional hearing loss due to occupational noise exposure. Examples of appropriate action include explanation of the effects of hearing loss, reinstruction and refitting of hearing protectors, additional training of the worker in hearing loss prevention, and reassignment of the worker to a quieter work area.

When the reviewing audiologist or physician suspects a hearing change is due to a non-occupational etiology, the worker shall receive appropriate counseling, which may include referral to his or her physician.

1.6.5 Exit Audiogram

The employer should obtain an exit audiogram from a worker who is leaving employment or whose job no longer involves exposure to hazardous noise. The exit audiogram should be conducted under the same conditions as those of baseline audiometry.

1.7 Hazard Communication

1.7.1 Warning Signs

A warning sign shall be clearly visible at the entrance to or the periphery of areas where noise exposures routinely equal or exceed 85 dBA as an 8-hr TWA. All warning signs shall be in English and, where applicable, in the predominant language of workers who do not read English. Workers unable to read the warning signs shall be informed verbally about the instructions printed on signs in hazardous work areas of the facility. The warning sign shall textually or graphically contain the following information:

```
WARNING

NOISE AREA
HEARING HAZARD

Use of Hearing Protectors Required
```

1.7.2 Notification to Workers

All workers who are exposed to noise at or above 85 dBA as an 8-hr TWA shall be informed about the potential consequences of noise exposure and the methods of preventing noise-induced hearing loss (NIHL). When noise measurements are initially conducted and confirm the presence of hazardous noise, or when followup noise measurements identify additional noise hazards, workers shall be notified within 30 days. New workers shall be alerted about the presence of hazardous noise before they are exposed to it.

1.8 Training

The employer shall institute a training program in occupational hearing loss prevention for all workers who are exposed to noise at or above 85 dBA as an 8-hr TWA; the employer shall ensure worker participation in such a program. The training program shall be repeated annually for each worker included in the HLPP. Information provided shall be updated to be consistent with changes in protective equipment and work processes.

The employer shall ensure that the training addresses, at a minimum, (1) the physical and psychological effects of noise and hearing loss; (2) hearing protector selection, fitting, use, and care; (3) audiometric testing; and (4) the roles and responsibilities of both employers and workers in preventing NIHL.

The format for the training program may vary from formal meetings to informal on-the-spot presentations. Allowances shall be made for one-on-one training, which would be particularly suitable for workers who have demonstrated a significant threshold shift. Whenever possible, the training should be timed to coincide with feedback on workers' hearing tests.

The employer shall maintain a record of educational and training programs for each worker for the duration of employment plus 1 year. On termination of employment, the employer should provide a copy of the training record to the worker. The employer may wish to keep the training record with the worker's exposure and medical records for longer durations (see Section 1.10).

1.9 Program Evaluation Criteria

The effectiveness of the HLPP shall be evaluated at the level of the individual worker and at the programmatic level.

The evaluation at the worker level shall take place at the time of the annual audiometry. If a worker demonstrates a significant threshold shift that is presumed to be occupationally related, all possible steps shall be taken to ensure that the worker does not incur additional occupational hearing loss.

The evaluation at the programmatic level shall take place annually. The incidence rate of significant threshold shift for noise-exposed workers shall be compared with that for a population not exposed to occupational noise. Similar incidence rates from this comparison indicate an effective HLPP. Data for calculating an incidence rate for a population not exposed to occupational noise should be drawn from Annex C in the *American National Standard Determination of Occupational Noise Exposure and Estimation of Noise-Induced Hearing Impairment*, ANSI S3.44–1996 [ANSI 1996c] unless more appropriate data are available.

1.10 Recordkeeping

The employer shall establish and maintain records in accordance with the requirements in Sections 1.10.1 through 1.10.5.

1.10.1 Exposure Assessment Records

The employer shall establish and maintain an accurate record of all exposure measurements required in Section 1.3. These records shall include, at a minimum, the name of the worker being monitored; identification number; duties performed and job locations; dates and times of measurements; type (refer to Section 6), brand, model, and size of hearing protectors used (if any); the measured exposure levels; and the identification of the person taking the measurements. Copies of a worker's exposure history resulting from this requirement shall also be included in the worker's medical file along with the worker's audiograms.

1.10.2 Medical Surveillance Records

The employer shall establish and maintain an accurate record for each worker subject to the medical surveillance specified in Section 1.6. These records shall include, at a minimum, the name of the worker being tested; identification number; duties performed and job locations; medical, employment, and noise-exposure history; dates, times, and types of tests (i.e., baseline, annual, retest, confirmation); hours since last noise exposure before each test; HTLs at the required audiometric frequencies; tester's identification and assessment of test reliability; the etiology of any significant threshold shift; and the identification of the reviewer.

1.10.3 Record Retention

In accordance with the requirements of 29 CFR[‡] 1910.20(d), Preservation of Records, the employer shall retain the records described in Sections 1.3 and 1.6 of this document for at least the following periods:

- 30 years for noise exposure monitoring records
- Duration of employment plus 30 years for medical monitoring records

In addition, records of audiometer calibrations and the ambient noise measurements in the audiometric testing room shall be maintained for 5 years.

1.10.4 Availability of Records

In accordance with 29 CFR 1910.20, Access to Employee Exposure and Medical Records, the employer shall, upon request, allow examination and provide copies of these records to a worker, a former worker, or anyone having appropriate authorization for record access.

1.10.5 Transfer of Records

The employer shall comply with the requirements for the transfer of records as set forth in 29 CFR 1910.20(h), Transfer of Records.

1.11 ANSI Standards

All standards (e.g., American National Standards Institute [ANSI] standards) referred to in this document shall be superseded by the latest available versions.

[‡]*Code of Federal Regulations.* See CFR in references.

Chapter 2

Introduction

2.1 Recognition of Noise as a Health Hazard

Noise, which is essentially any unwanted or undesirable sound, is not a new hazard. Indeed, NIHL has been observed for centuries. Before the industrial revolution, however, comparatively few people were exposed to high levels of workplace noise. The advent of steam power in connection with the industrial revolution first brought general attention to noise as an occupational hazard. Workers who fabricated steam boilers developed hearing loss in such numbers that the malady was dubbed "boilermaker's disease." Increasing mechanization in all industries and most trades has since proliferated the noise problem.

2.2 Noise-Induced Hearing Loss (NIHL)

NIHL is caused by exposure to sound levels or durations that damage the hair cells of the cochlea. Initially, the noise exposure may cause a temporary threshold shift—that is, a decrease in hearing sensitivity that typically returns to its former level within a few minutes to a few hours. Repeated exposures lead to a permanent threshold shift, which is an irreversible sensorineural hearing loss.

Hearing loss has causes other than occupational noise exposure. Hearing loss caused by exposure to nonoccupational noise is collectively called sociocusis. It includes recreational and environmental noises (e.g., loud music, guns, power tools, and household appliances) that affect the ear the same as occupational noise. Combined exposures to noise and certain physical or chemical agents (e.g., vibration, organic solvents, carbon monoxide, ototoxic drugs, and certain metals) appear to have synergistic effects on hearing loss [Hamernik and Henderson 1976; Brown et al. 1978; Gannon et al. 1979; Brown et al. 1980; Hamernik et al. 1980; Pryor et al. 1983; Rebert et al. 1983; Humes 1984; Boettcher et al. 1987; Young et al. 1987; Byrne et al. 1988; Fechter et al. 1988; Johnson et al. 1988; Morata et al. 1993; Franks and Morata 1996]. Some sensorineural hearing loss occurs naturally because of aging; this loss is called presbycusis. Conductive hearing losses, as opposed to sensorineural hearing losses, are usually traceable to diseases of the outer and middle ear.

Noise exposure is also associated with nonauditory effects such as psychological stress and disruption of job performance [Cohen 1973; EPA 1973; Taylor 1984; Öhrström et al. 1988; Suter 1989] and possibly hypertension [Parvizpoor 1976; Jonsson and Hansson 1977; Takala et al. 1977; Lees and Roberts 1979; Malchaire and Mullier 1979;

Manninen and Aro 1979; Singh et al. 1982; Belli et al. 1984; Delin 1984; Talbott et al. 1985; Verbeek et al. 1987; Wu et al. 1987; Talbott et al. 1990]. Noise may also be a contributing factor in industrial accidents [Cohen 1976; Schmidt et al. 1980; Wilkins and Acton 1982; Moll van Charante and Mulder 1990]. Nevertheless, data are insufficient to endorse specific damage risk criteria for these nonauditory effects.

2.3 Physical Properties of Sound

The effects of sound on a person depend on three physical characteristics of sound: amplitude, frequency, and duration. Sound pressure level (SPL), expressed in decibels, is a measure of the amplitude of the pressure change that produces sound. This amplitude is perceived by the listener as loudness. In sound-measuring instruments, weighting networks (described in Chapter 4) are used to modify the SPL. Exposure limits are commonly measured in dBA. When used without a weighted network suffix, the expression should be dB SPL.

The frequency of a sound, expressed in Hz, represents the number of cycles occurring in 1 sec and determines the pitch perceived by the listener. Humans with normal hearing can hear a frequency range of about 20 Hz to 20 kilohertz (kHz). Exposures to frequency ranges that are considered infrasonic (below 20 Hz), upper sonic (10 to 20 kHz), and ultrasonic (above 20 kHz) are not addressed in this document.

Although no uniformly standard definitions exist, noise exposure durations can be broadly classified as continuous-type or impulsive. All nonimpulsive noises (i.e., continuous, varying, and intermittent) are collectively referred to as "continuous-type noise." Impact and impulse noises are collectively referred to as "impulsive noise." Impulsive noise is distinguished from continuous-type noise by a steep rise in the sound level to a high peak followed by a rapid decay. In many workplaces, the exposures are often a mixture of continuous-type and impulsive sounds.

2.4 Number of Noise-Exposed Workers in the United States

In 1981, OSHA estimated that 7.9 million U.S. workers in the manufacturing sector were occupationally exposed to daily noise levels at or above 80 dBA [46 Fed. Reg.* 4078 (1981a)]. In the same year, the U.S. Environmental Protection Agency (EPA) estimated that more than 9 million U.S. workers were occupationally exposed to daily noise levels above 85 dBA, as follows:

---

*Federal Register. See Fed. Reg. in references.

Major group	Number of workers
Agriculture	323,000
Mining	255,000
Construction	513,000
Manufacturing and utilities	5,124,000
Transportation	1,934,000
Military	976,000
Total	9,125,000

More than half of these workers were engaged in manufacturing and utilities [EPA 1981].

From 1981 to 1983, NIOSH conducted the National Occupational Exposure Survey (NOES), which was designed to provide data describing the occupational safety and health conditions in the United States [NIOSH 1988a,b, 1990]. The surveyors visited and gathered information at various workplaces throughout the United States. For the purposes of NOES, workers were considered noise-exposed if any noise (excluding impulsive noise) at or above 85 dBA occurred in their work environment at least once per week for 90% of the workweeks in a year [NIOSH 1988a]. Because not all industries were surveyed, NOES does not provide an all-inclusive estimate of the number of noise-exposed workers in the United States; however, it does provide reasonable estimates of the numbers of noise-exposed workers in the particular industries covered by NOES. These estimates are tabulated in Table 2-1, which shows that noise-exposed workers were employed in a wide range of industries, with the majority in manufacturing.

To collect occupational health data in mining industries not covered by NOES, NIOSH conducted the National Occupational Health Survey of Mining (NOHSM) from 1984 to 1989. Unlike NOES surveyors, the NOHSM surveyors did not measure the noise levels but used qualitative evaluation to determine noise exposures. As shown in Table 2-2, noise exposures occurred in all of the industries covered by NOHSM.

2.5 Legislative History

Efforts to regulate occupational noise in the United States began about 1955. The military was first to establish such regulations for members of the Armed Forces [U.S. Air Force 1956]. Under the Walsh-Healey Public Contracts Act of 1936, as amended, safety and health standards had been issued that contained references to excessive noise; however, they prescribed neither limits nor acknowledged the occupational hearing loss problem. A later regulation under this act [41 CFR 50-204.10], promulgated in 1969, defined noise limits that were applicable only to those firms having supply contracts with the U.S. Government greater than $10,000; similar limits were made applicable to work under Federal service contracts of $2,500 or more under the Service Contract Act. The noise rule in the Walsh-Healey Act regulations was adopted under the Federal Coal Mine Health and Safety Act of 1969 (Public Law 91-173) for underground and surface coal mine operations.

Table 2–1. Estimated number of workers exposed to noise at or above 85 dBA, by economic sector (two-digit SIC*,†)

Economic sector	SIC	Total number of production workers	Noise-exposed production workers Number	As % of total production workers
Agriculture, forestry, and fishing:				
Agriculture services	07	89,189	17,618	19.8
Mining:				
Oil and gas extraction	13	330,841	76,525	23.1
Construction:				
General building contractors	15	664,833	105,299	15.8
Heavy construction, except building	16	517,969	124,610	24.0
Special trade contractors	17	1,228,744	191,087	15.6
Manufacturing:				
Food and kindred products	20	1,188,267	343,030	28.9
Tobacco products	21	106,399	57,764	54.3
Textile mill products	22	615,322	262,108	42.6
Apparel and other finished products	23	1,082,236	150,824	13.9
Lumber and wood products	24	475,730	196,489	41.3
Furniture and fixtures	25	428,539	121,271	28.3
Paper and allied products	26	488,101	164,808	33.8
Printing and publishing	27	724,707	154,862	21.4
Chemicals and allied products	28	592,059	102,671	17.3
Petroleum and coal products	29	160,516	31,998	19.9
Rubber and miscellaneous plastics products	30	595,525	135,611	22.8
Leather and leather products	31	144,200	9,346	6.5
Stone, clay, and glass products	32	457,983	98,215	21.5
Primary metal industries	33	824,725	269,270	32.7
Fabricated metal products	34	1,151,777	336,919	29.3
Industrial machinery and equipment	35	1,544,883	229,509	14.9
Electronic and other electric equipment	36	1,287,842	104,553	8.1
Transportation equipment	37	1,311,750	238,609	18.2
Instruments and related products	38	555,108	48,014	8.7
Miscellaneous manufacturing industries	39	418,805	39,307	9.4

See footnotes at end of table. (Continued)

Table 2-1 (Continued). Estimated number of workers exposed to noise at or above 85 dBA, by economic sector (two-digit SIC*,†)

Economic sector	SIC	Total number of production workers	Noise-exposed production workers Number	As % of total production workers
Transportation and public utilities:				
Local and inter-urban passenger transit	41	171,428	14,832	8.7
Trucking and warehousing	42	561,058	39,150	7.0
Transportation by air	45	312,931	94,656	30.3
Communications	48	387,505	23,124	6.0
Electric, gas, and sanitary services	49	588,041	89,730	15.3
Wholesale trade:				
Wholesale trade—durable goods	50	528,659	110,283	20.9
Wholesale trade—nondurable goods	51	99,410	5,287	5.3
Retail trade:				
Automotive dealers and service stations	55	334,063	4,543	1.4
Services:				
Personal services	72	366,545	33,462	9.1
Business services	73	766,108	11,246	1.5
Auto repair, services, and parking	75	320,459	33,997	10.6
Miscellaneous repair services	76	143,302	12,682	8.9
Health services	80	2,679,610	15,677	0.6
Total		24,245,169	4,098,986	16.9

*Standard industrial classification. Source: OMB [1987].
†Based on data collected by NOES [NIOSH 1988a,b, 1990]. Not all two-digit SIC sectors and not all four-digit SIC industries within each two-digit SIC sector were surveyed. The NOES covered 39 of 83 two-digit SIC sectors, and the NOES estimates were representative of only the four-digit SIC industries actually surveyed. For example, within agricultural services (SIC 07), the estimates are for crop preparation services (SIC 0723), veterinary services for animal specialties (SIC 0742), lawn and garden services (SIC 0782), and ornamental shrub and tree services (SIC 0783) only, because no surveys were done for soil preparation services (SIC 0711), crop planting and protecting (SIC 0721), crop harvesting (SIC 0722), cotton ginning (SIC 0724), veterinary services for livestock (SIC 0741), livestock services (SIC 0751), animal specialty services (SIC 0752), farm labor contractors (SIC 0761), farm management services (SIC 0762), and landscape counseling and planning (SIC 0781).

Table 2-2. Estimated number of workers exposed to noise, by industry (four-digit SIC*)†

Industry	SIC	Total number of production workers	Noise-exposed production workers Number	As % of total production workers
Iron ores	1011	3,614	3,411	94.4
Copper ores	1021	8,777	8,253	94.0
Lead and zinc ores	1031	1,363	1,190	87.3
Gold ores	1041	3,574	3,041	85.1
Silver ores	1044	1,893	1,503	79.4
Ferroalloy ores, except vanadium	1061	713	653	91.6
Uranium-radium-vanadium ores	1094	1,177	952	80.9
Miscellaneous metal ores, not elsewhere classified	1099	3,798	3,322	87.5
Bituminous coal and lignite mining	1220	123,274	108,264	87.8
Anthracite mining	1231	2,006	1,704	85.0
Crude petroleum and natural gas‡	1311	107	101	94.4
Dimension stone	1411	2,122	1,837	86.6
Crushed and broken limestone	1422	26,906	19,292	71.7
Crushed and broken granite	1423	4,545	3,643	80.2
Crushed and broken stone, not elsewhere classified	1429	5,796	4,829	83.3
Sand and gravel	1440	13,825	11,519	83.3
Clay, ceramic, and refractory minerals	1459	8,171	6,829	83.6
Potash, soda, and borate minerals	1474	4,855	4,258	87.7
Phosphate rock	1475	4,422	3,209	72.6
Chemical and fertilizer minerals	1479	2,175	1,297	59.6
Miscellaneous nonmetallic minerals	1499	4,755	3,586	75.4
Chemical preparation, not elsewhere classified‡	2899	263	250	95.1
Petroleum and coal products, not elsewhere classified‡	2999	42	23	54.8
Cement, hydraulic‡	3241	5,681	4,757	83.7
Lime‡	3274	2,529	2,014	79.6
Total	—	236,383	199,737	84.5

*Standard industrial classification. Source: OMB [1987].
†Based on data collected by NOHSM (unpublished data).
‡Estimates apply only to the miners—not the total workforce in this SIC industry.

In 1970, the Occupational Safety and Health Act (Public Law 95–164) was enacted, which established OSHA within the U.S. Department of Labor as the enforcement agency responsible for protecting the safety and health of a large segment of the U.S. workforce. Concurrently, NIOSH was established under the Department of Health, Education, and Welfare (now the Department of Health and Human Services) to develop criteria for safe occupational exposures to workplace hazards. In compliance with this provision, NIOSH published *Criteria for a Recommended Standard: Occupational Exposure to Noise* in 1972 [NIOSH 1972]. The document provided the basis for a recommended standard to reduce the risk of developing permanent noise-induced occupational hearing loss. The criteria document presented an REL of 85 dBA as an 8-hr TWA and methods for measuring noise, calculating noise exposure, and providing a hearing conservation program. However, the criteria document acknowledged that (1) NIOSH was not able to determine the technical feasibility of the REL, and (2) approximately 15% of the population exposed to occupational noise at the 85-dBA level for a working lifetime would develop occupational NIHL.

Initially, OSHA adopted the Walsh-Healey exposure limit of 90 dBA as an 8-hr TWA with a 5-dB exchange rate as its permissible exposure limit (PEL) [29 CFR 1910.95] for general industry. In 1974, responding to the NIOSH criteria document, OSHA proposed a revised noise standard [39 Fed. Reg. 37773 (1974a)] but left the PEL unchanged. The proposed standard was not promulgated; however, it articulated the requirement for a hearing conservation program. In 1981 and again in 1983, OSHA amended its noise standard to include specific provisions of a hearing conservation program for occupational exposures at 85 dBA or above [46 Fed. Reg. 4078 (1981a); 48 Fed. Reg. 9738 (1983)]. The OSHA noise standard as amended does not cover all industries. For example, the Hearing Conservation Amendments do not cover noise-exposed workers in transportation, oil/gas well drilling and servicing, agriculture, construction, and mining. The construction industry is covered by another OSHA noise standard [29 CFR 1926.52]; the mining industry is regulated by four separate standards that are enforced by MSHA [30 CFR 56.5050; 30 CFR 57.5050; 30 CFR 70.500–70.508; 30 CFR 71.800–71.805]. These standards vary in specific requirements regarding exposure monitoring and hearing conservation; however, all maintain an exposure limit based on 90 dBA for an 8-hr duration. Although they are required to comply with OSHA regulations by Executive Order 12196, the U.S. Air Force [1993] and the U.S. Army [1994] have chosen a more stringent exposure limit of 85 dBA as an 8-hr TWA with a 3-dB exchange rate. Thus, the protection that a worker receives from occupational noise depends in part on the sector in which he or she is employed.

The exposure limits discussed above apply only to continuous-type noises. For impulsive noise, the generally accepted limit not to be exceeded for any time is a peak level of 140 dB SPL. Among the regulatory standards, this peak level is either enforceable or nonenforceable, as indicated by the word "shall" or "should," respectively. For example, in the MSHA standards for metal and nonmetal mines [30 CFR 56.5050; 30 CFR 57.5050], this exposure limit is enforceable; in the OSHA standards [29 CFR 1910.95; 29 CFR 1926.52], it is nonenforceable.

2.6 Scope of This Revision of the Noise Criteria Document

The focus of this document is on the prevention of occupational hearing loss rather than on conservation. Prevention means to avoid creating hearing loss. Conservation means to sustain the hearing that is present, regardless of whether damage has already occurred. An emphasis on prevention evolves from beliefs that it should not be necessary to suffer an impairment, illness, or injury to earn a living and that it is possible to use methods to prevent occupational hearing loss. This document evaluates and presents recommended exposure limits, a 3-dB exchange rate, and other elements necessary for an effective HLPP. Where the information is incomplete to support definitive recommendations, research needs are suggested for future criteria development. Nonauditory effects of noise and hearing losses due to causes other than noise are beyond the scope of this document.

CHAPTER 3

Basis for the Exposure Standard

3.1 Quantitative Risk Assessment

The selection of an exposure limit depends on the definitions of two parameters: (1) the maximum acceptable occupational hearing loss (i.e., the fence) and (2) the percentage of the occupational noise-exposed population for which the maximum acceptable occupational hearing loss will be tolerated. The fence is often defined as the average HTL for two, three, or four audiometric frequencies. It separates the maximum acceptable hearing loss from smaller degrees of hearing loss and normal hearing. Excess risk is the difference between the percentage that exceeds the fence in an occupational-noise-exposed population and the percentage that exceeds it in an unexposed population. Mathematical models are used to describe the relationship between excess risk and various factors such as average daily noise exposure, duration of exposure, and age group.

The most common protection goal is the preservation of hearing for speech discrimination. Using this protection goal, NIOSH [1972] employed the term "hearing impairment" to define its criteria for maximum acceptable hearing loss; and OSHA later used the slightly modified term "material hearing impairment" to define the same criteria [46 Fed. Reg. 4078 (1981a)]. In this context, a worker was considered to have a material hearing impairment when his or her average HTLs for *both* ears exceeded 25 dB at the audiometric frequencies of 1000, 2000, and 3000 Hz (denoted here as the "1-2-3-kHz definition").

3.1.1 NIOSH Risk Assessment in 1972

NIOSH [1972] assessed the *excess* risk of material hearing impairment as a function of levels and durations (e.g., 40-year working lifetime) of occupational noise exposure. Thus, for a 40-year lifetime exposure in the workplace to average daily noise levels of 80, 85, or 90 dBA, the excess risk of material hearing impairment was estimated to be 3%, 16%, or 29%, respectively. On the basis of this risk assessment, NIOSH recommended an 8-hr TWA exposure limit of 85 dBA [NIOSH 1972].

To compare the NIOSH excess risk estimates with those developed by other organizations, the NIOSH data were also analyzed using the same 25-dB fence, but averaging the HTLs at 500, 1000, and 2000 Hz (the 0.5-1-2-kHz definition) [NIOSH 1972]. Table 3-1 presents the excess risk estimates developed by NIOSH [1972], EPA [1973], and the International Standards Organization (ISO) [1971] for material hearing impairment caused by occupational noise exposure. OSHA used these estimates as the basis for requiring hearing conservation programs for occupational noise exposures at or above 85 dBA (8-hr TWA) [46 Fed. Reg. 4078 (1981a)].

Table 3–1. Estimated excess risk of incurring material hearing impairment[*] as a function of average daily noise exposure over a 40-year working lifetime[†]

Reporting organization	Average daily noise exposure (dBA)	Excess risk (%)[‡]
ISO	90	21
	85	10
	80	0
EPA	90	22
	85	12
	80	5
NIOSH	90	29
	85	15
	80	3

[*]For purposes of comparison in this table, material hearing impairment is defined as an average of the HTLs for both ears at 500, 1000, and 2000 Hz that exceeds 25 dB.
[†]Adapted from 39 Fed. Reg. 43802 [1974b].
[‡]Percentage with material hearing impairment in an occupational-noise-exposed population after subtracting the percentage who would normally incur such impairment from other causes in an unexposed population.

The data used for the NIOSH risk assessment were collected by NIOSH in 13 noise and hearing surveys (collectively known as the Occupational Noise and Hearing Survey [ONHS]) from 1968 to 1971. The industries in the surveys included steelmaking, paper bag processing, aluminum processing, quarrying, printing, tunnel traffic controlling, woodworking, and trucking. Questionnaires and audiometric examinations were given to noise-exposed and non-noise-exposed workers who had consented to participate in the surveys. More than 4,000 audiograms were collected, but the sample excluded audiograms of (1) noise-exposed workers whose noise exposures could not be characterized relative to a specified continuous noise level over their working lifetime, and (2) noise-exposed workers with abnormal hearing levels as determined by their medical history. Thus, 1,172 audiograms were used. These represented 792 noise-exposed and 380 non-noise-exposed workers (controls) [NIOSH 1972; Lempert and Henderson 1973].

3.1.2 NIOSH Risk Assessment in 1997

A review of relevant epidemiologic literature did not identify new data suitable for estimating the excess risk of occupational NIHL for U.S. workers. The prolific use of hearing protectors in the U.S. workplace since the early 1980's would confound determination of dose-response relationships for occupational NIHL among contemporary workers. Therefore, the current risk assessment is based on a reanalysis of data from the NIOSH ONHS [Prince et al. 1997].

Prince et al. [1997] (reprinted in the Appendix of this document) derived a new set of excess risk estimates using the ONHS data with a model referred to as the "1997-NIOSH model," which differed from the 1972-NIOSH model [NIOSH 1972]. A noteworthy difference between the two models is that Prince et al. [1997] considered the possibility of nonlinear effects of noise in the 1997-NIOSH model, whereas the 1972-NIOSH model was based solely on a linear assumption for the effects of noise. Table 3–2 provides an overview of the differences between the 1997- and the 1972-NIOSH models. Prince et al. [1997] found that nonlinear models fit the data well and that the linear models similar to the 1972-NIOSH model did not fit as well. In addition to using the 0.5-1-2-kHz and the 1-2-3-kHz definitions of material hearing impairment to assess the risk of occupational NIHL, Prince et al. [1997] used the definition of hearing handicap* proposed by the American Speech-Language-Hearing Association (ASHA) Task Force on the Definition of Hearing Handicap. Prince et al. [1997] found the ASHA Task Force definition† (average of HTLs at 1000, 2000, 3000, and 4000 Hz) [ASHA 1981] useful because it was geared toward excess risk of hearing loss rather than compensation. Phaneuf et al. [1985] also found that the audiometric average of 1000, 2000, 3000, and 4000 Hz provided "a superior prediction of hearing disability in terms of specificity, sensitivity, and overall accuracy." The ASHA Task Force definition is also referred to as the 1-2-3-4-kHz definition in this criteria document. Table 3–3 presents the excess risk estimates for this definition and associated 95% confidence intervals.

The ISO has also developed procedures for estimating hearing loss due to noise exposure. In 1971, the ISO issued the first edition of *ISO 1999, Assessment of Occupational Noise Exposure for Hearing Conservation Purposes* [ISO 1971] (referred to as the "1971-ISO model"), which included risk estimates for material hearing impairment from occupational noise exposures. In 1990, the ISO issued a second edition of *ISO 1999, Acoustics—Determination of Occupational Noise Exposure and Estimation of Noise-Induced Hearing Impairment* [ISO 1990] (referred to as the "1990-ISO model"). Both ISO models are based on broadband, steady noise exposures for 8-hr work shifts during a working lifetime of up to 40 years.

The various models for estimating the excess risk of material hearing impairment are compared in Table 3–4. The excess risk estimates derived from the 1971-ISO, 1972-NIOSH, 1973-EPA, and 1997-NIOSH‡ models are reasonably similar. However,

---

*ASHA makes a distinction between the terms "impairment" and "handicap"; however, for the purpose of the subsequent discussion in this criteria document, only the term "material hearing impairment" is used. The Prince et al. [1997] paper reports the use of a modified ASHA Task Force definition. This modification incorporates frequency-specific weights based on the articulation index for each frequency [ANSI 1969]. Negligible differences were found between excess risk estimates generated using the modified and the unmodified definitions. The excess risk estimates presented in this criteria document are based on the unmodified ASHA Task Force definition.

†Historical note, ASHA did not deliberate on the definition proposed by the ASHA Task Force.

‡Prince et al. [1997] found that the excess risk estimates at exposure levels below 85 dBA were not well defined. Insufficient data for workers with average daily exposures below 85 dBA led to considerable variability in the estimation, depending on the statistical assumptions used in the modeling.

Table 3–2. Comparison of the 1997– and 1972–NIOSH risk-damage models

Item	Description	
	1997–NIOSH model	1972–NIOSH model
Model	Logit model: Dichotomous outcome[*] Model probability of hearing impairment directly	Probit model: Continuous outcome (average HTL) Model distribution of HTL and calculate percentage of population meeting impairment criteria
Sound level effect	Dependent on duration of exposure $\beta [L_e-L_0]^\phi$ L_0 (control sound level) and ϕ (shape of dose-response curve) are estimated from the data L_e=Sound level in exposed population Model allows flexibility in determining shape of dose-response curve and location of control sound levels	Dependent on duration of exposure $\beta [L_e-L_0]^1$ L_0 and ϕ are fixed values $\phi=1$ assumes a linear dose-response relationship L_e=Sound level in exposed population
Age, years	Modeled as a continuous variable	Modeled as a categorical variable with five levels (17–27, 28–35, 36–45, 46–54, 55–70)
Duration of exposure, years	Modeled as a categorical variable with 4 levels (<2, 2–4, 5–10, >10)	Modeled as a categorical variable with five levels (<2, 2–4, 5–10, 11–20, 21–41)

[*]Each individual is categorized either as hearing-impaired (defined as average HTL >25 dB, both ears) or non-hearing-impaired (average HTL ≤25 dB).

except for the 1-2-3-4-kHz definition, the excess risk estimates derived from the 1990-ISO model are considerably lower than those derived from the other models. These disparities may be due to differences in the statistical methodology or in the underlying data used. Nevertheless, these five models confirm an excess risk of material hearing impairment at 85 dBA.

As mentioned earlier in this section, the protection goal incorporated in the definitions of material hearing impairment has been to preserve hearing for speech discrimination. The 4000-Hz audiometric frequency is recognized as being both sensitive to noise and important for hearing and understanding speech in unfavorable or noisy listening conditions [Kuzniarz 1973; Aniansson 1974; Suter 1978; Smoorenburg 1990]. In recognition of the fact that listening conditions are not always ideal in everyday life, and in concurrence with the ASHA [1981] Task Force proposal, NIOSH has modified its

Chapter 3. Basis for the Exposure Standard

Table 3–3. Excess risk estimates for material hearing impairment,* by age and duration of exposure

Average daily exposure (dbA)	5–10 years of exposure								>10 years of exposure							
	Age 30		Age 40		Age 50		Age 60		Age 30		Age 40		Age 50		Age 60	
	Risk (%)	95% CI†	Risk (%)	95% CI	Risk (%)	95% CI	Risk (%)	95% CI	Risk (%)	95% CI	Risk (%)	95% CI	Risk (%)	95% CI	Risk (%)	95% CI
90	5.4	2.1–9.5	9.7	3.7–16.5	14.3	5.5–24.4	15.9	6.2–26.2	10.3	5.8–16.2	17.5	10.7–25.3	24.1	14.6–33.5	24.7	14.9–34.3
85	1.4	0.3–3.2	2.6	0.6–6.0	4.0	0.9–9.3	4.9	1.0–11.5	2.3	0.7–5.3	4.3	1.3–9.4	6.7	2.0–13.9	7.9	2.3–16.6
80	0.2	0–1.1	0.4	0–2.2	0.6	0.01–3.6	0.8	0.01–4.7	0.3	0–1.8	0.6	0.01–3.3	1.0	0.01–5.2	1.3	0.01–6.8

*1997–NIOSH model for the 1-2-3-4-kHz definition of hearing impairment.
†CI=confidence interval.

Table 3–4. Comparison of models for estimating the excess risk of
material hearing impairment at age 60 after a 40-year working lifetime
exposure to occupational noise, by definition of material hearing impairment

Average exposure level (dBA)	0.5–1–2-kHz definition					1–2–3-kHz definition			1–2–3–4-kHz definition	
	1971-ISO	1972-NIOSH	1973-EPA	1990-ISO	1997-NIOSH	1972-NIOSH	1990-ISO	1997-NIOSH	1990-ISO	1997-NIOSH
90	21	29	22	3	23	29	14	32	17	25
85	10	15	12	1	10	16	4	14	6	8
80	0	3	5	0	4	3	0	5	1	1

definition of material hearing impairment to include 4000-Hz when assessing the risk of occupational NIHL. Therefore, with this modification, NIOSH defines material hearing impairment as an average of the HTLs for both ears that exceeds 25 dB at 1000, 2000, 3000, and 4000 Hz. Based on this definition, the excess risk is 8% for workers exposed to an average daily noise level of 85 dBA over a 40-year working lifetime. NIOSH continues to recommend the REL of 85 dBA as an 8-hr TWA on the basis of (1) analyses supporting the 1972 REL of 85 dBA as an 8-hr TWA, (2) reanalyses of the ONHS data, (3) ASHA Task Force positions on preservation of speech discrimination, and (4) analyses of excess risk of ISO, EPA, and NIOSH databases.

For extended work shifts (i.e., greater than 8 hr), lower exposure limits can be extrapolated from the REL of 85 dBA as an 8-hr TWA (see Section 1.1.1 or Table 1–1). Stephenson et al. [1980] studied human responses to 24-hr noise exposures and found that no temporary threshold shift occurred for broadband noise exposures less than 75 to 80 dBA. These data are in line with the recommendation that TWA exposures be less than 80 to 81 dBA for durations greater than 16 hr.

3.2 Ceiling Limit

Because NIOSH is recommending a 3-dB exchange rate with an 85-dBA REL, a ceiling limit for continuous-type noise is unnecessary. For example, with an 85-dBA REL and a 3-dB exchange rate, an exposure duration of less than 28 sec would be allowed at a 115-dBA level.

The generally accepted ceiling limit of 140 dB peak SPL for impulsive noise is based on a report by Kryter et al. [1966]. Ward [1986] indicated that "this number was little more than a guess when it was first proposed." To date, a proposal for a different limit has not been supported. Henderson et al. [1991] indicated that the critical level for chinchillas is between 119 and 125 dB; and if a 20-dB adjustment is used to account for the difference in susceptibility between chinchillas and humans, the critical level extrapolated for humans would be between 139 and 145 dB. Based on the 85-dBA REL and the 3-dB exchange rate, the allowable exposure time at 140 dBA is less than 0.1 sec; thus, 140 dBA is a reasonable ceiling limit for impulsive noise.

3.3 Exchange Rate

Health effects depend on exposure level and duration. The NIOSH recommendation for a 3-dB exchange rate is based in part on the conclusions from a NIOSH contract report [Suter 1992a]. This report involved an exhaustive analysis of the relationship between hearing loss, noise level, and exposure duration. Although the time/intensity relationship is most commonly referred to as the exchange rate, it is also referred to as the "doubling rate," "trading ratio," and "time-intensity tradeoff." The 3-dB exchange rate is also known as the equal-energy rule or hypothesis, because a 3-dB increase/decrease represents a doubling or halving of the sound energy. The most commonly used exchange rates incorporate either 3 dB or 5 dB per doubling or halving of exposure duration [Embleton 1994].

The 3-dB exchange rate is the method most firmly supported by scientific evidence for assessing hearing impairment as a function of noise level and duration. This rate is already used in the United States by the EPA and the U.S. Department of Defense. The 3-dB exchange rate is used worldwide by nations such as Canada, Australia, New Zealand, the People's Republic of China, the United Kingdom, Germany, and many others. First proposed by Eldred et al. [1955], the 3-dB exchange rate was later supported by Burns and Robinson [1970]. The premise behind the 3-dB exchange rate is that equal amounts of sound energy will produce equal amounts of hearing impairment regardless of how the sound energy is distributed in time. Theoretically, this principle could apply to exposures ranging from a few minutes to many years. However, Ward and Turner [1982] suggest restricting its use to the sound energy accumulated in 1 day. They distinguish between (1) an interpretation of the total energy theory that would allow an entire lifetime of exposure to be condensed into a few hours and (2) a restricted equal-A-weighted-daily-energy interpretation of the theory. Burns [1976] also cautions against the misuse of the equal-energy hypothesis, noting that it was based on data gathered from workers who experienced 8-hr occupational exposures daily for periods of months to years; thus, extrapolation to very different conditions would be inappropriate.

In 1973, the U.S. Air Force adopted a 4-dB exchange rate [U.S. Air Force 1973]. This exchange rate is based on an unpublished analysis by H.O. Parrack at the Aerospace Medical Research Laboratory. However, a set of curves based on this analysis was published as Figure 20 in a joint EPA/Air Force report [Johnson 1973]. The 4-dB exchange rate came closest to the curve that best described temporary theshold shift at 1000-Hz audiometric frequency [Johnson 1973]. However, Johnson [1973] also pointed out that according to these curves, the 3-dB exchange rate would best protect hearing at the 4000-Hz frequency, and the 5-dB exchange rate would be a good compromise if hearing were to be protected only at the midfrequencies—500, 1000, and 2000 Hz.

The relationship between the 3-dB exchange rate and energy can be illustrated as follows. The *American National Standard for Acoustical Terminology*, ANSI S1.1-1994 [ANSI 1994] defines the decibel as a "unit of level when the base of the logarithm is the tenth root of ten, and the quantities concerned are proportional to power.... [E]xamples of quantities that qualify are power (in any form), sound pressure squared, particle velocity squared, sound intensity, sound-energy density, and voltage squared. Thus, the decibel is a unit of sound-pressure-squared level; it is common practice, however, to shorten this to sound pressure level, when no ambiguity results from so doing."

Ostergaard [1986] provided a functional elucidation of the relationships pointed to in the ANSI definition:

> In acoustics, decibel notation is utilized for most quantities. The *decibel* is a dimensionless unit based on the logarithm of the ratio of a measured quantity to a reference quantity. Thus, decibels are defined as follows:
>
> $$L = k \log_{10} (A/B)$$
>
> where L is the level in decibels, A and B are quantities having the same units, and k is a multiplier, either 10 or 20 depending on whether A and B are measures of energy or pressure, respectively. Any time a level is referred to in acoustics, decibel notation is implied. In acoustics all *levels* are referred to some reverence quantity, which is the denominator, B, of the equation.

Applying this mathematical relationship in the following calculations demonstrates how every doubling of energy yields an increase of 3 dB:

Let X = the exchange rate whereby energy is doubled
$10 \log_{10} (A/B) + X = 10 \log_{10} (2A/B)$
$X = 10 \log_{10} (2A/B) - 10 \log_{10} (A/B)$
$\quad = 10 \log_{10} (2)$
$\quad = 10 (0.301)$
$\quad = 3.01$ dB

This same relationship does not hold true for the 5-dB exchange rate. To derive $X = 5$ dB, the sound intensity would have to be more than doubled in this equation. Thus, the 5-dB exchange rate does not provide for the doubling or halving of energy per 5-dB increment.

The 5-dB exchange rate is sometimes called the OSHA rule; it is less protective than the equal-energy hypothesis. The 5-dB exchange rate attempts to account for the interruptions in noise exposures that commonly occur during the workday [40 Fed. Reg. 12336 (1975)], presuming that some recovery from temporary threshold shift occurs during these interruptions and the hearing loss is not as great as it would be if the noise were continuous. The rule makes no distinction between continuous and noncontinuous noise, and it will permit comparatively long exposures to continuous noise at higher sound levels than would be allowed by the 3-dB rule. On the basis of the limited data that existed in the early 1970's, NIOSH [1972] recommended the 5-dB exchange rate; however, after reviewing the more recent scientific evidence, NIOSH now recommends the 3-dB exchange rate.

The evolution of the 5-dB exchange rate began in 1965 when the Committee on Hearing, Bioacoustics, and Biomechanics (CHABA) for the National Academy of Sciences—National Research Council issued criteria for assessing allowable exposures to continuous, fluctuating, and intermittent noise [Kryter et al. 1966]. The CHABA criteria were an attempt to predict the hazard from nearly every conceivable noise exposure pattern based on temporary threshold shift experimentation. In the development of its criteria, CHABA used the following postulates:

1. TTS_2 (temporary threshold shift measured 2 min after a period of noise exposure) is a consistent measure of the effects of a single day of exposure to noise.

2. All noise exposures that produce a given TTS_2 will be equally hazardous (the equal temporary effect theory).

3. Permanent threshold shift produced after many years of habitual noise exposures for 8 hr per day is about the same as the TTS_2 produced in normal ears by an 8-hr exposure to the same noise.

However, these CHABA postulates were not validated. Research has been unable to demonstrate a simple relationship between temporary threshold shift, permanent threshold shift, and cochlear damage [Burns and Robinson 1970; Ward 1970, 1980; Ward and Turner 1982; Hétu 1982; Clark and Bohne 1978, 1986]. The CHABA criteria assumed that worker exposures could be characterized by regularly spaced noise bursts interspersed with periods that were sufficiently quiet to allow hearing to recover. However, this assumption is not characteristic of many typical industrial noise exposures. Workers will always develop temporary threshold shift before sustaining permanent threshold shift, barring an ototraumatic incident. Temporary threshold shift is a useful metric for monitoring the effects of noise exposure; these studies do not imply otherwise.

In general, the CHABA hearing damage risk criteria proved too complicated for general use. Botsford [1967] published a simplified set of criteria based on the CHABA criteria. One of the simplifications inherent to the Botsford [1967] method was the assumption that interruptions would be of "equal length and spacing so that a number of identical exposure cycles would be distributed uniformly throughout the day." These interruptions would occur during coffee breaks, trips to the washroom, lunch, and periods when machines were temporarily shut down.

During the same period, another related development led to the 5-dB exchange rate. Simplifying the criteria developed by Glorig et al. [1961] and adopted by ISO [1961], the Intersociety Committee [1970] published its criteria, which consisted of a table showing permissible exposure levels (starting at 90 dBA) as a function of duration and the number of occurrences per day. The exchange rates varied considerably depending on noise level and frequency of occurrence. For continuous noise with durations of less than 8 hr, the Committee recommended maximum exposure levels based on a 5-dB exchange rate. The only field study that has been repeatedly cited as supporting the 5-dB rule is one study of coal miners by Sataloff et al. [1969].

In 1969, the U.S. Department of Labor promulgated a noise standard [34 Fed. Reg. 790 (1969a)] under the authority of the Walsh-Healey Public Contracts Act. The standard contained a PEL of 90 dBA for continuous noise. Exposure to varying or intermittent noise was to be assessed over a weekly period according to a large table of exposure indices. The exchange rate varied according to level and duration: a rate of 2 to 3 dB was used for long-duration noises of moderate level, and 6 to 7 dB was used for short-duration, high-level bursts. This standard was withdrawn after a short period. Later in 1969, the Walsh-Healey noise standard that is in effect today was issued [34 Fed. Reg. 7948 (1969b)]. In this version, any special criteria for varying or intermittent noise had disappeared, and the 5-dB exchange rate became official. Thus, the 5-dB exchange rate appears to have been the outgrowth of the many simplifying processes that preceded it.

Beginning with the study of Burns and Robinson [1970], the credibility of the 3-dB rule has been increasingly supported by numerous studies and by national and international consensus [EPA 1973, 1974; 39 Fed. Reg. 43802 (1974b); ISO 1971; von Gierke et al. 1981; ISO 1990; U.S. Air Force 1993; U.S. Army 1994; ACGIH 1995].

Data from a number of field studies correspond well to the 3-dB rule (equal-energy hypothesis), as Passchier-Vermeer [1971, 1973] and Shaw [1985] have demonstrated. In Passchier-Vermeer's [1973] portrayal of the data, the Passchier-Vermeer [1968] and the Burns and Robinson [1970] prediction models for hearing losses as a function of continuous-noise exposure level fit the data on hearing losses from varying or intermittent noise exposures quite well. The fact that comparisons using the newer ISO standard [ISO 1990] corroborate Passchier-Vermeer's findings lend additional support to the equal-energy hypothesis.

Some older field data from occupations such as forestry and mining show less hearing loss than expected when compared with equivalent levels of continuous noise [Sataloff et al. 1969; Holmgren et al. 1971; Johansson et al. 1973; INRS 1978]. However, these findings have not been supported by the two NIOSH [1976, 1982] studies of intermittently exposed workers or the analyses conducted by Passchier-Vermeer [1973] and Shaw [1985].

Data from animal experiments support the use of the 3-dB exchange rate for single exposures of various levels within an 8-hr day [Ward and Nelson 1971; Ward and Turner

1982; Ward et al. 1983]. Nevertheless, several animal studies have demonstrated that some recovery may occur during the "quiet" periods of an intermittent noise exposure [Bohne and Pearse 1982; Ward and Turner 1982; Ward et al. 1982; Bohne et al. 1985; Bohne et al. 1987; Clark et al. 1987]. However, these benefits are likely to be smaller or even nonexistent in the industrial environment, where sound levels during quiet periods are considerably higher and where interruptions are not evenly spaced.

The possible ameliorative effect of intermittency does not justify the use of the 5-dB exchange rate. For example, although Ward [1970] noted that some industrial studies have shown lower permanent threshold shifts from intermittent noise exposure than would be predicted by the 3-dB rule, he did not favor selection of the 5-dB exchange rate as a compromise to compensate for the effects of intermittency, because it would allow single exposures at excessively high levels. In his opinion, "this compromise was futile and perhaps even dangerous" [Ward 1970].

One response to the evidence from the animal studies and certain field studies would be to select the 3-dB exchange rate but to allow an adjustment (increase) to the PEL for certain intermittent noise exposures, as suggested by EPA [1974] and Johansson et al. [1973]. This response would be in contrast to a 5-dB exchange rate, for which there is little scientific justification. Ideally, if an adjustment is needed, the amount should be determined by the temporal pattern of the noise and the levels of quiet between noise bursts. At this time, however, little quantitative information is available about these parameters in industrial environments. Therefore, the need for an adjustment should be clarified by further research. Although the 3-dB rule may be somewhat conservative in truly intermittent conditions, the 5-dB rule will be underprotective in most others. The 3-dB exchange rate is the method most firmly supported by the scientific evidence for assessing hearing impairment as a function of noise level and duration, whether or not an adjustment is used for certain intermittent exposures.

3.4 Impulsive Noise

The OSHA occupational noise standard [29 CFR 1910.95] states: "Exposure to impulsive or impact noise should not exceed 140 dB peak sound pressure." Thus, in this context, the 140-dB limit is advisory rather than mandatory. This number was first proposed by Kryter et al. [1966] and later acknowledged by Ward [1986] as little more than a guess. NIOSH [1972] did not address the hazard of impulsive (i.e., impulse or impact) noise, although NIOSH stated that the provisions of the recommended standard in the criteria document were intended to apply for all noise. Although there is yet no unanimity as to which criteria best describe the relationship between NIHL and exposure to impulsive noise, either by itself or in the presence of continuous-type (i.e., continuous, varying, or intermittent) noise, there is an international standard that has become widely used by most industrial nations. This standard, *ISO 1999, Acoustics—An Estimation of Noise-Induced Hearing Impairment* [ISO 1990], integrates both impulsive and continuous-type noise (and uses the 3-dB exchange rate of the equal-energy rule) when calculating sound exposures over any specified time period. NIOSH concurs with this approach and recommends that noise exposure levels be calculated by integrating all noises (both impulsive and continuous-type) over the duration of the measurement.

Despite its simplicity, the equal-energy rule is not universally accepted as a method for characterizing exposures that consist of both impulsive and continuous-type noises. Another approach favors evaluating impulsive noise separate from that of continuous-type noise. Studies that would argue for this approach will be discussed first, followed by a discussion of studies elucidating the rationale for the NIOSH position on the equal-energy rule.

3.4.1 Evidence That Impulsive Noise Effects Do Not Conform to the Equal-Energy Rule

In her evaluation of the effects of continuous and varying noises on hearing, Passchier-Vermeer [1971] found that the HTLs of workers in steel construction works did not conform to the equal-energy hypothesis; that is, the hearing losses in these workers, who were exposed to noise levels with impulsive components, were higher than predicted. Later studies by Ceypek et al. [1973], Hamernik and Henderson [1976], and Nilsson et al. [1977] also indicated that continuous and impulsive noises have a synergistic rather than additive effect on hearing.

Comparing the studies of Passchier-Vermeer [1973] and of Burns and Robinson [1970], Henderson and Hamernik [1986] suggested that the steeper slope of Passchier-Vermeer's exposure-response curve at the 4000-Hz audiometric frequency might have been due to noise exposures that contained impulsive components, a characteristic not present in the Burns and Robinson data. Citing the similarity of Passchier-Vermeer's data to those collected by Taylor et al. [1984] and Kuzniarz et al. [1976] on workers exposed to impulsive noise environments, Henderson and Hamernik [1986] indicated that exposure to continuous and impulsive noises in combination may be more hazardous than exposure to continuous noise alone.

Voight et al. [1980] studied noise exposure patterns in the building construction industry and related the equivalent continuous sound level for 8 hr (L_{Aeq8hr}) to audiometric records of more than 81,000 construction workers in Sweden. They found differences in hearing loss among groups exposed to noise of the same L_{Aeq8hr} but with different temporal characteristics. Groups exposed to impulsive noise had more hearing loss than those exposed to continuous noise of the same L_{Aeq8hr}.

Sulkowski and Lipowczan [1982] conducted noise measurement and audiometric testing in a drop-forge factory. The HTLs of 424 workers in the factory were compared with the predicted values according to the Burns and Robinson equation [1970]. The observed and predicted values differed in that the observed hearing loss was smaller than predicted at the lower audiometric frequencies, but the observed hearing loss was greater than predicted at the higher audiometric frequencies. In their study of hearing loss in weavers, who were exposed to continuous noise, and drop-forge hammer men, who were exposed to impact noise of equivalent energy, Sulkowski et al. [1983] found that the hammer men had substantially worse hearing than the weavers.

Thiery and Meyer-Bisch [1988] conducted a cross-sectional epidemiologic study at an automobile manufacturing plant. The automotive workers were exposed to continuous and impulsive noises at L_{Aeq8hr} ranging from 87 to 90 dBA. When their HTLs were compared with those of workers exposed to continuous noise at L_{Aeq8hr} of 95 dBA for the same exposure time, the automotive workers showed greater hearing losses at the 6000-Hz audiometric frequency than the reference population after 9 years of exposure.

Starck et al. [1988] compared at the 4000-Hz audiometric frequency the HTLs of forest workers using chain saws and shipyard workers using hammers and chippers. The forest workers were exposed to continuous-type noise, whereas the shipyard workers were exposed to impact noise. Starck et al. [1988] also used the immission model developed by Burns and Robinson [1970] to predict the HTLs for both groups. They found that the Burns and Robinson model was accurate at 4000 Hz for the forest workers; however, it substantially underestimated the HTLs at 4000 Hz for the shipyard workers.

The studies described here provide evidence that the effects of combined exposure to impulsive and continuous-type noises are synergistic rather than additive, as the equal-energy hypothesis would support. One measure for protecting a worker from such synergistic effects would be to require that a correction factor be added to a measured TWA noise exposure level when impulsive components are present in the noise. The magnitude of such a correction has not been quantified. The matter becomes more complicated when other parameters of impulsive noise are considered. Noise energy does not appear to be the only factor that affects hearing. The amplitude, duration, rise time, number of impulses, repetition rate, and crest factor also appear to be involved [Henderson and Hamernik 1986; Starck and Pekkarinen 1987; Pekkarinen 1989]. The criteria for exposure to impulsive noise based on the interrelationships of these parameters await the results of further research.

3.4.2 Evidence That Impulsive Noise Effects Conform to the Equal-Energy Rule

In 1968, CHABA published damage risk criteria for impulsive noise based on the equal-energy hypothesis [Ward 1968]. Over the years, individuals and organizations have supported treating impulsive noise on an equal-energy basis [Coles et al. 1973; EPA 1974; Coles 1980; ISO 1990].

Burns and Robinson [1970] proposed the concept of immission, which is based on the equal-energy hypothesis, to describe the total energy from a worker's exposure to continuous noise over a period of time (i.e., months or years). Atherley and Martin [1971] modified this concept to include impulsive noise in the calculation of the L_{Aeq8hr}.

In a study of 76 men who were exposed to impact noise in two drop-forging factories, Atherley and Martin [1971] calculated each man's noise exposure (immission level) during his employment period and plotted it against his age-corrected HTLs over six audiometric frequencies. They found that the observed HTLs of the population came close to the predicted HTLs according to Robinson [1968] and concluded that the equal-energy hypothesis was applicable to impact noise. Similarly, Atherley [1973] examined the HTLs of 50 men exposed to impact noise produced by pneumatic chisels used on metal castings and found good agreement between observed and predicted HTLs.

Guberan et al. [1971] compared the HTLs of 70 workers exposed to impact noise in drop-forging workshops with the predicted HTLs according to Robinson [1968] at the 3-, 4-, and 6-kHz audiometric frequencies. Again, the observed HTLs were in close agreement with the predicted HTLs.

A study of 716 hammer and press operators in 7 drop forges by Taylor et al. [1984] indicated that hearing losses resulting from impact and continuous noises in the drop-forging industry were as great or greater than those resulting from equivalent continuous noise. Using noise dosimetry, Taylor et al. [1984] found that the hammer operators were exposed to an average L_{Aeq8h} of 108 dBA, whereas the press operators were exposed to 99 dBA. The investigators also conducted audiometry for the operators. The median HTLs of hammer operators of all age groups approximated those predicted by the Robinson [1968] immission model. The median HTLs of younger press operators (aged 15 to 34) also corresponded closely with the predicted values; however, those of older press operators (aged 34 to 54) were significantly higher than predicted. These results indicate that, up to certain limits, the equal-energy hypothesis can be applied to combined exposure to impact and continuous noises.

3.4.3 Combined Exposure to Impulsive and Continuous-Type Noises

In many industrial operations, impulsive noise occurs in concert with a background of continuous-type noise. In some animal studies the effects of combined exposure to continuous-type and impulsive noises appear to be synergistic at high exposure levels [Hamernik et al. 1974]. But the synergism disappears when the exposure levels are comparable with those found in many common industrial environments [Hamernik et al. 1981]. Whether the effects of combined exposure are additive or synergistic, exposure to these noises causes hearing loss; thus the contribution of impulse noise to the noise dose should not be ignored. If the effects are additive, the 85-dBA REL with the 3-dB exchange rate would be sufficiently protective. If the effects are synergistic, the same would still be protective to a smaller extent. NIOSH therefore recommends that the REL of 85 dBA as an 8-hr TWA be applicable to all noise exposures, whether such exposures are from continuous-type noise, impulsive noise, or combined continuous-type and impulsive noises.

CHAPTER 4

Instrumentation for Noise Measurement

No single method or process exists for measuring occupational noise. Hearing safety and health professionals can use a variety of instruments to measure noise and can choose from a variety of instruments and software to analyze their measurements. The choice of a particular instrument and approach for measuring and analyzing occupational noise depends on many factors, not the least of which will be the purpose for the measurement and the environment in which the measurement will be made. In general, measurement methods should conform to the *American National Standard Measurement of Occupational Noise Exposure*, ANSI S12.19–1997 [ANSI 1996a]. However, it is beyond the scope of this document to serve as a manual for operating equipment and making sound measurements. Rather, this chapter will be limited to concise remarks relevant to operating the two most commonly used instruments for measuring noise exposures: the sound level meter and the noise dosimeter. More detailed discussions about instrumentation and measurement protocols appear in reference sources such as NIOSH [1973], Earshen [1986], Johnson et al. [1991], and Harris [1991].

4.1 Sound Level Meter

The sound level meter is the basic measuring instrument for noise exposures. It consists of a microphone, a frequency selective amplifier, and an indicator. At a minimum, it measures sound level in dB SPL. An integrating function may be included to automate the calculation of the TWA or the noise dose.

4.1.1 Frequency Weighting Networks

The human ear is not equally responsive to all frequencies; it is most sensitive around 4000 Hz and least sensitive in the low frequencies. The responses of the sound level meter are modified with frequency-weighting networks that represent some responses of the human ear. These empirically derived networks approximate the equal loudness-weighting networks or scales; some also have a B-scale. The A-scale, which approximates the ear's response to moderate-level sounds, is commonly used in measuring noise to evaluate its effect on humans and has been incorporated in many occupational noise standards. Table 4–1 shows the characteristics of these scales.

4.1.2 Exponential Time Weighting

A sound level meter's response is generally based on either a FAST or SLOW exponential averaging. FAST corresponds to a 125-millisecond (ms) time constant; SLOW

Noise Exposure

Table 4–1. Relative response of sound level meter weighting networks[*]

Octave-center frequency (Hz)	Weighted response (dB)		
	A scale	B scale	C scale
31.5	-39.4	-17.1	-3.0
63	-26.2	-9.3	-0.8
125	-16.1	-4.2	-0.2
250	-8.6	-1.3	0
500	-3.2	-0.3	0
1,000	0	0	0
2,000	1.2	-0.1	-0.2
4,000	1.0	-0.7	-0.8
8,000	-1.1	-2.9	-3.0
16,000	-6.6	-8.4	-8.5

[*]Adapted from ANSI [1983].

corresponds to a 1-s time constant. The meter dynamics are such that the meter will reach 63% of the final steady-state reading within one time constant. The meter indicator reflects the average SPL measured by the meter during the period selected. In most industrial settings, the meter fluctuates less when measurements are made with the SLOW response compared with the FAST response. A rapidly fluctuating sound generally yields higher maximum SPLs when measured with a FAST response. The choice of meter response depends on the type of noise being measured, the intended use of the measurements, and the specifications of any applicable standard. For typical occupational noise measurements, NIOSH recommends that the meter response on a sound level meter be set at SLOW.[*]

4.1.3 Microphones for Sound Level Meters

The correct use of the microphone is extremely important in obtaining accurate measurements. Microphones come in many types and sizes. A microphone is typically designed for use in a particular environment across a specific range of SPLs and frequencies. In addition, microphones differ in their directionality. For example, some are intended to be pointed directly at the sound; and others are designed to measure sound from a "grazing" angle of incidence. Thus users should follow the sound level meter manufacturer's instructions regarding the type and size of microphone and its orientation toward a sound. Also, care should be taken to avoid shielding the microphone by persons or objects [ANSI 1996a]. When measuring a diffuse sound field, the person conducting the measurement should hold the microphone as far from his or her body as practical [Earshen 1986].

---

[*]Meters that are set to integrate or average sound do not use either the FAST or SLOW time constant; they will sample many times each second. For a more detailed description of exponential time weighting, refer to Yeager and March [1991].

4.2 Noise Dosimeter

Measuring noise with a sound level meter is relatively simple when the noise levels are continuous and when the worker remains essentially stationary during the work shift. A noise dosimeter is preferred for measuring a worker's noise exposure when the noise levels are varying or intermittent, when they contain impulsive components, or when the worker moves around frequently during the work shift.

The noise dosimeter may be thought of as a sound level meter with an additional storage and computational function. It measures and stores the sound levels during an exposure period and computes the readout as the percent dose or TWA. Many dosimeters available today can provide an output in dose or TWA using various exchange rates (e.g., 3, 4, and 5 dB), 8-hr criterion levels (e.g., 80, 84, 85, and 90 dBA), and sound measurement ranges (e.g., 80 to 130 dBA). The choice of FAST or SLOW meter response on the dosimeter does not affect the computed noise dose or TWA when the 3-dB exchange rate is used, but it will when other exchange rates are used [Earshen 1986].

In noise dosimetry, the microphone is attached on the worker whose exposure is being measured. The placement of the microphone is important in estimating the worker's exposure, as Kuhn and Guernsey [1983] have found large differences in the sound distribution about the body. ANSI [1996a] specifies that the microphone be located on the midtop of the worker's more exposed shoulder and that it be oriented approximately parallel to the plane of this shoulder.

4.3 Range of Sound Levels

OSHA requires that, for the purposes of the Hearing Conservation Amendment, all sound levels from 80 to 130 dBA be included in the noise measurements [29 CFR 1910.95(d)(2)(I)]. This range was specified on the basis of instrument capabilities available at that time [ANSI 1978], and OSHA had intended to increase the upper limit of the range to 140 or 150 dB as improved dosimeters became readily available [46 Fed. Reg. 4135 (1981b)].

To measure all sound levels from 80 to 140 dBA, a noise dosimeter should have an operating range of at least 63 dB and a pulse range of the same magnitude. In contrast, the ANSI S1.25-1991 standard specifies that dosimeters should have an operating range of at least 50 dB and a pulse range of at least 53 dB [ANSI 1991a]. Today, noise dosimeters with operating and pulse ranges in excess of 65 dB are quite common. Therefore, NIOSH considers that measuring all sound levels from 80 to 140 dBA with a noise dosimeter is technically feasible.

CHAPTER 5

Hearing Loss Prevention Programs (HLPPs)

Whenever hazardous noise exists in the workplace, measures should be taken to reduce noise levels as much as possible to protect exposed workers and to monitor the effectiveness of these intervention processes. Employers have an obligation to protect their workers from this debilitating occupational hazard [46 Fed. Reg. 4078 (1981a); 48 Fed. Reg. 9738 (1983)]. In addition, research has shown that implementing effective HLPPs (also known as hearing conservation programs) has numerous other benefits in the workplace [NIOSH 1996]. For example, Cohen [1976] found reduced employee absenteeism following the establishment of a hearing conservation program. Similarly, Schmidt et al. [1980] reported a reduction in workplace injuries following the introduction of a hearing conservation program. Alternatively, other reports have documented detrimental nonauditory effects of noise, such as decreased productivity in high noise environments [Noweir 1984; Suter 1992b]. Employers who effectively protect their workers' hearing may also reap the economic benefits of lower workers' compensation rates because of fewer claims for NIHL.

NIOSH recommends that HLPPs be implemented for all workers whose unprotected 8-hr TWA exposures (i.e., exposures incurred without the use of hearing protectors) equal or exceed 85 dBA and that the programs include at least the following components [NIOSH 1996]:

1. Initial and annual audits of procedures

2. Assessment of noise exposures

3. Engineering or administrative control of noise exposures

4. Audiometric evaluation and monitoring of workers' hearing

5. Use of hearing protectors for exposures equal to or greater than 85 dBA, regardless of exposure duration

6. Education and motivation of workers

7. Recordkeeping

8. Program evaluation for effectiveness

Today, no legitimate reason exists for any worker to incur an occupational hearing loss [NIOSH 1996]. Implementation of an HLPP must hinge on the fact that occupational NIHL is 100% preventable. The key to developing and implementing an effective program lies in a commitment by both management and workers to prevent hearing loss [Helmkamp et al. 1984]. This end is facilitated by integrating the HLPP into the company's overall health and safety program [Berger 1981; NIOSH 1996]. This step gives the prevention of hearing loss the same weight as the prevention of other work-related illnesses and injuries, thus indicating to workers and management that occupational hearing loss must be taken seriously. Other factors that facilitate an effective HLPP include encouraging workers to carry over their good hearing conservation practices to off-the-job situations; using simple, clearly-defined procedures; making compliance with the HLPP a condition of employment; and incorporating safety requirements into written company policy.

5.1 Personnel Requirements

Responsibility for developing and implementing an HLPP usually resides with a team of professionals. The American Occupational Medical Association (AOMA) [1987] identifies the team approach to hearing conservation as necessary for its success. The number of team members and their professional disciplines may vary with the kind of company and the number of noise-exposed workers; however, members frequently include audiologists, physicians, occupational health nurses, occupational hearing conservationists, engineers, industrial hygienists, safety professionals, management representatives, and employee and union safety representatives.

Regardless of whether program responsibility resides with a team or a single individual, one person should act as champion for the program, maintaining overall responsibility for its implementation [NIOSH 1996; Royster and Royster 1990]. This individual will be referred to in this document as the "program implementor." The program implementor should ensure that all aspects of the program are fully and properly administered and should enlist the support of management and workers in actively preventing hearing loss. Royster and Royster [1990] recommend that the primary qualification of the program implementor be a genuine interest in preserving workers' hearing. AOMA [1987] recommends that the program implementor be a physician. NIOSH [1996] maintains that the professional discipline of the program implementor is not as important as his or her ability to act as the champion of the HLPP by focusing management and worker attention on hearing conservation issues. In addition, the program implementor's stature in the organization should allow him or her to make decisions, correct deficiencies, enforce compliance, and supervise other team members with regard to the program.

In addition to the program implementor, one person should be responsible for the audiometric aspects of the HLPP; this person will be referred to in this document as the "audiometric manager." The professional qualifications of this person are critical. The audiometric manager should be an audiologist or a physician specializing in otological or occupational medicine. The program implementor and the audiometric manager may be the same person—provided that he or she meets the qualifications for both positions. If the program implementor and the audiometric manager are not the same person, the audiometric manager should report to the program implementor, regardless of the professional credentials of either party.

5.2 Initial and Annual Audits (Component 1)

Ideally, an initial audit should be conducted before an HLPP is implemented or any changes are made to an existing program. This audit will serve as a basis for assessing the effectiveness of an improved program. The audit should begin by examining administrative issues such as corporate responses to safety and health regulations, official policies promoting good safety and health practices, assurance of adequate resources to conduct the program, and the status of the program implementor within the company. Current engineering and administrative controls should be evaluated, and the systems for monitoring noise exposures and conducting audiometry should be critically examined. Employee and management training should be noted, and past successes and failures should be analyzed so that improvements can be made. In particular, if engineering and administrative controls are insufficient, auditors should note whether effective training is provided in the selection, fitting, and daily use of hearing protectors. Recordkeeping procedures should be inspected meticulously because methods for maintaining records of audiometry, noise exposure, and other aspects of the overall program can greatly influence the success or failure of a program. NIOSH recommends that an HLPP audit be conducted annually as a part of an overall program evaluation so that the strengths of the program may be clearly identified and weaknesses promptly addressed [NIOSH 1996].

5.3 Exposure Assessment (Component 2)

Section 6(b)(7) of the Occupational Safety and Health Act of 1970 [29 USC 651 et seq.] requires that, where appropriate, occupational health standards provide for monitoring or measuring employee exposure at the locations and intervals and in the manner necessary for the protection of employees. Accurate characterization of the noise hazard present in the workplace and the subsequent identification of affected workers are both extremely important. These two elements form the basis for all subsequent actions within the HLPP [NIOSH 1996]. Monitoring procedures should be specifically defined to ensure consistency. Instrumentation, calibration, measurement parameters, and methods for linking results to worker records should be clearly delineated. Exposure assessment should be done during typical production cycles; however, if noise levels vary significantly during different phases of production, then exposures should be assessed separately for each phase [Royster and Royster 1990; NIOSH 1996].

Exposure assessment should be conducted by an industrial hygienist, audiologist, or other professional with appropriate training [NIOSH 1996]. Workers should be permitted and encouraged to observe and participate in monitoring activities insofar as such observation or participation does not interfere with the monitoring procedure. Their participation will help ensure valid results, as the workers frequently have the experience to identify the prevailing noise sources, indicate periods when noise exposure may differ, and recognize whether given noise levels are typical or atypical. They can explain how different operating modes affect equipment sound levels and they can describe worker tasks and positions. The cooperation of workers is also critical to ensure that workers do not advertently or inadvertently interfere with obtaining valid measurements. The initial exposure monitoring can serve as an introduction to the HLPP by raising the awareness of workers and management regarding noise as a hazard. The monitoring survey, if conducted cooperatively, can help establish a rapport that will help obtain the cooperation of both workers and essential management in later phases of the program [Royster and Royster 1990; NIOSH 1996].

The frequency with which noise exposure assessments are updated depends on several variables. These might include the intensity of the noise, potential changes in exposure due to changes in equipment or production, the rate of significant threshold shift noted among workers, other changes noted in additional measures of program effectiveness, requirements of various governmental regulations, workers' compensation requirements of individual States, union contract stipulations, and specific company policies [Royster et al. 1986].

In general, after the initial exposure assessment, NIOSH [1996] recommends that exposure monitoring be repeated periodically—at least every 2 years for noise levels equal to or greater than 95 dBA and at least every 5 years for noise levels less than 95 dBA. Periodic noise monitoring will identify situations where the noise levels have changed because of, for example, aging equipment, equipment with maintenance problems, and undocumented process changes. Monitoring shall be repeated sooner if a change in production, process, equipment, or personnel might affect exposure levels [Royster et al. 1986; Royster and Royster 1990; NIOSH 1996].

Workers shall be notified of the noise exposure level determined for their particular job and the relative risk that such an exposure poses to their hearing. This information should also be cross-referenced to individual worker records. Notification should include a description of the specific hazardous noise sources in the worker's area, the purpose and proper use of any noise control devices, and requirements for hearing protectors, if appropriate. This notification can be incorporated into the worker training program [Royster and Royster 1990; NIOSH 1996]. The notification may also be posted in the work area. Noise contour maps may be posted and readily available for the entire facility, so that workers may be made aware of the noise levels in other areas. In cases where noise is due to a process, notification may include a list of noise-hazardous processes.

At a minimum, warning signs should be posted on the periphery of noise areas [Royster and Royster 1990; NIOSH 1996]. The warning signs should include a requirement that hearing protectors be worn in the area, and a supply of several types of hearing protectors should be readily accessible. Signs should communicate to workers graphically and should be printed in English and in the predominant language of the workers who do not read English.

5.4 Engineering and Administrative Controls (Component 3)

For occupational hearing loss prevention, NIOSH defines *engineering control* as "any modification or replacement of equipment, or related physical change at the noise source or along the transmission path (with the exception of hearing protectors) that reduces the noise level at the employee's ear" [NIOSH 1996]. Typical mechanisms for engineering noise controls include reducing noise at the source (installing a muffler), altering the noise path (building an acoustic enclosure or barrier), reducing reverberation (covering walls with sound-absorbing materials), and reducing equipment vibration (installing vibration mounts). Engineering controls should be the first order of protection from excessive noise exposure [46 Fed. Reg. 4078 (1981a); Suter 1986; AOMA 1987]. When the noise can be reduced or eliminated through engineering controls, the danger to hearing is also reduced or eliminated. Where periodic noise monitoring is conducted, the feasibility of employing engineering controls should be reevaluated, with priority given to noise sources that affect the greatest number of workers. *Any* reduction in noise level (even if it is only a few decibels) serves to make the noise hazard more manageable, reduces the risk of hearing loss, improves communication, and lowers annoyance and related extra-auditory problems associated with high noise levels [NIOSH 1996]. Furthermore, when the noise can be reduced to acceptable levels through engineering controls, employers may forego some of the additional difficulties and expenses related to providing hearing protectors, education and motivation programs, and program evaluation [Royster and Royster 1990].

To reduce noise in an existing facility, it is generally necessary to retrofit engineering controls. Development of these controls should involve engineers, safety and industrial hygiene personnel, and the workers who operate, service, and maintain the equipment. Development of special noise control measures must be predicated on a thorough assessment of the noise source and individual worker exposure. Consideration should be given to the relative contribution of each noise source to the overall sound levels. Various noise control options should be evaluated on the basis of their effectiveness, cost, technical feasibility, and implications for equipment use, service, and maintenance. Other potential complications of new noise control measures (such as effects on lighting, heat production, ventilation, and ergonomics) should be considered [NIOSH 1996]. Engineering controls must always consider the proper maintenance of equipment. In addition, the function and purpose of any planned or existing engineering controls should be fully discussed with the workers so that they support the controls and do not inadvertently interfere with them [NIOSH 1996].

Management should also consider noise reduction when planning for new or remodeled facilities. Engineering controls can be most effective when they are incorporated into the design and purchase of equipment from the start. In addition, the cost of incorporating engineering controls during the design phase is generally much lower than retrofitting them at a later date. The ultimate noise level can be substantially reduced by substituting more sound-absorbent materials, modifying equipment structure or mechanical processes, and isolating sources within the equipment [Haag 1988a].

A "buy-quiet" policy for new equipment acquisitions should be adopted by management [Royster and Royster 1990; Brogan and Anderson 1994; NIOSH 1996]. Haag [1988b] describes a four-part process that management can implement to have an effective buy-quiet policy. The process includes selecting products or operations to be targeted for noise reduction through new purchases, setting criteria for new equipment noise levels, requesting noise level specifications from manufacturers, and including these noise level data in bid evaluation. Again, input from workers should be incorporated into the buying process.

When engineering controls are inadequate, supplemental administrative controls may be utilized to help limit exposures. *Administrative controls* are defined as changes in the work schedule or operations that reduce worker noise exposures. For example, sometimes workers can be scheduled so that their time in a noisy environment is minimized. When extremely noisy operations are unavoidable, the number of workers permitted to work in such an environment should be minimized. In all cases, the application of administrative controls should not result in exposing more workers to noise. Finally, a quiet, clean, and conveniently located lunch and break area should be provided to give workers periodic relief from workplace noise.

5.5 Audiometric Evaluation and Monitoring (Component 4)

Audiometric evaluation of workers' hearing is crucial to the success of an HLPP because it is the only way to actually determine whether occupational hearing loss is being prevented. Because occupational hearing loss occurs gradually, affected employees often notice no change in hearing ability until a relatively large change in their hearing sensitivity has occurred. The annual comparison of audiometric tests can trigger prompt hearing loss program interventions, initiating protective measures and motivating employees to prevent further hearing loss.

5.5.1 Audiometry

Audiometry shall be conducted by an audiologist, a physician, or by an occupational hearing conservationist certified by the CAOHC or the equivalent. All testing shall be supervised by an audiologist, an otologist, or an occupational physician. Occupational hearing conservationists should follow the training guidelines proposed by the National Hearing Conservation Association (NHCA) [1987]. Use of microprocessor-based or self-recording audiometers should not waive the qualification requirements for the tester.

For audiometric testing to be beneficial, management must allocate sufficient time and resources to allow for timely and accurate testing. The testing must be conducted carefully to ensure the integrity of the audiometric data. Effective communication and coordination are critical among management, health service providers, and workers.

Audiometry shall, at a minimum, consist of pure-tone air-conduction threshold testing of each ear at 500, 1000, 2000, 3000, 4000, and 6000 Hz. Although this entire frequency range is not used in the assessment of OSHA's standard threshold shift (STS), all of these frequencies are important in deciding the probable etiology of a hearing loss. To enhance the decision about probable etiology, testing at 8000 Hz should also be considered. Sufficient time should be taken to conduct the test accurately. Testing too quickly sacrifices accuracy and gives the worker the impression that audiometry and the HLPP are unimportant [NIOSH 1996].

Audiograms are displayed and stored as tables or charts of hearing thresholds measured in each ear at specified test frequencies. In OSHA-mandated hearing conservation programs, thresholds must be measured for pure-tone signals at the test frequencies of 500, 1000, 2000, 3000, 4000, and 6000 Hz [29 CFR 1910.95(h)(1)]. At each frequency, the threshold recorded for an ear is the lowest signal output level of the audiometer at which the individual responds in a specified percentage of trials (such as 50%) or in two of three trials. Hearing thresholds are measured in dB HTL (decibels, hearing threshold level), with 0 dB HTL representing average hearing ability for young people with no otological pathology.* Larger threshold values indicate poorer-than-average hearing; smaller threshold values (negative thresholds such as -5 or -10 dB) indicate better than average hearing.

A person's audiometric threshold at a given test frequency is not an unchanging quantity. Measurement variability is associated with the state of the subject (including the subject's prior audiometric experience, attention, motivation, the influence of upper respiratory problems, drugs, and other factors) and with the testing equipment and methodology [Morrill 1986]. The higher the measurement variability, the more difficult it is to distinguish actual changes in hearing threshold.

Noise exposure increases hearing thresholds, resulting in threshold shifts toward higher values (poorer hearing). Occasionally, exposure to extremely intense noise may cause an immediate, permanent hearing loss known as acoustic trauma. Most often, exposure to less intense noise causes the gradual development of hearing damage over months and years. During each overexposure to noise the ear develops a temporary reduction in sensitivity called temporary threshold shift. This shift reverses over a period of hours or days if the ear is allowed to rest in a quieter environment. However, if the exposure is high enough or if exposures are repeated, the temporary threshold shift may not reverse completely, and a permanent threshold shift begins to develop.

Although the magnitude of the temporary threshold shift cannot be used to predict the magnitude of the permanent threshold shift, the former serves as a precursor to the latter.

---

*Whenever the unit dB is used in audiometric testing, it actually refers to dB HTL.

NIOSH therefore suggests that monitoring audiometry be conducted on noise-exposed workers at the end of or late in their daily work shifts. Discovering and taking action to prevent further temporary threshold shift will result in more thorough worker protection from permanent hearing damage. If the annual monitoring audiometry is performed at the beginning of work shifts or before the workday begins, temporary threshold shifts that might have been present from the previous day's noise exposure will have been resolved—any threshold shifts observed will represent permanent shifts in hearing. This type of audiometric monitoring will serve only to document the development of permanent hearing loss, not to prevent it.

Some reports have indicated that industrial audiometry is too variable to be useful in detecting initial threshold shifts [Hétu 1979; Atherley and Johnston 1981]. Certainly, if testing procedures are too inconsistent, temporary or permanent threshold shifts may not be distinguishable from measurement variability. The challenge is to select a criterion for significant threshold shift that is stringent enough to detect incipient hearing loss, yet not so stringent as to identify large numbers of workers whose thresholds are simply showing normal variability. This challenge is compounded by the fact that the incipient permanent threshold shift may manifest itself with the same order of magnitude as typical audiometric measurement variability—about a 10-dB change in hearing thresholds. However, the daily temporary threshold shift is often larger in magnitude than the developing permanent threshold shift. So testing workers near the end of their work shifts (when temporary threshold shifts may be present) should increase the probability of identifying workers who are not adequately protected from noise.

In 1972, a significant threshold shift criterion was initially recommended by NIOSH [NIOSH 1972]. In 1992 and 1996, Royster [1992, 1996] examined the performance of this criterion against seven other criteria for significant threshold shift. The following threshold shift criteria were evaluated:

1. OSHA STS: in either ear, a change of 10 dB or more in the average of hearing thresholds at 2000, 3000, and 4000 Hz.

2. OSHA STS TWICE: in either ear, a change of 10 dB or more in the average of hearing thresholds at 2000, 3000, and 4000 Hz is present on one annual audiogram and is persistent in the same ear on the next audiogram.

3. American Academy of Otolaryngology—Head and Neck Surgery (AAO-HNS) SHIFT: in either ear, a change of 10 dB or more in the average of hearing thresholds at 500, 1000, and 2000 Hz, or 15 dB or more at 3000, 4000, and 6000 Hz.

4. 1972 NIOSH SHIFT: in either ear, a change of 10 dB or more at 500, 1000, 2000, or 3000 Hz, or 15 dB or more at 4000 or 6000 Hz.

5. 15-dB SHIFT: in either ear, a change of 15 dB or more at any test frequency from 500 through 6000 Hz.

6. 15-dB TWICE: in either ear, a change of 15 dB or more at any test frequency from 500 through 6000 Hz is present on one annual audiogram and is persistent at the same frequency in the same ear on the next audiogram.

7. 15-dB TWICE 1–4 kHz: in either ear, a change of 15 dB or more at any test frequency from 1000 through 4000 Hz is present on one annual audiogram and is persistent at the same frequency in the same ear on the next audiogram.

8. 10-dB AVG 3–4 kHz: in either ear, a change of 10 dB or more in the average of hearing thresholds at 3000 and 4000 Hz.

The study methodology, database characteristics, and results are described in detail in the Royster [1992, 1996] reports. This study compared each of the above eight criteria for threshold shifts by applying each criterion to 15 different industrial hearing conservation databases that were contributed to the ANSI S12 Working Group 12.

Within each database, analyses were restricted to the first eight audiograms for male workers who had at least eight tests. The numbers of workers included from each database ranged from 39 to 1,056. Data were analyzed for a total of 2,903 workers. For the purposes of these analyses, a "tag" was identified when a worker's audiogram (or two consecutive audiograms for the TWICE criteria) met a specified criterion, and a "true positive" was identified when the worker's audiogram showed the same threshold shift specified in that criterion.

A significant threshold shift for a worker, according to the four nonaveraging, any-frequency-shift criteria (1972 NIOSH SHIFT, 15-dB SHIFT, 15-dB TWICE, and 15-dB TWICE 1–4 kHz), was considered a true positive if the shift was confirmed by the succeeding audiogram—but only if the shift was persistent for at least one of the same frequencies in the same ear. For example, if a worker's Test 3 showed a 1972 NIOSH SHIFT at 2000, 4000, and 6000 Hz in the left ear, then the shift would be confirmed as a true positive if Test 4 showed the shift to be persistent in the same ear at one or more of the same frequencies. For three of the frequency-average criteria (OSHA STS, AAO-HNS SHIFT, and 10-dB AVG 3–4 kHz), a shift was considered a true positive if the worker's next audiogram showed a change by that same criterion, whether or not the confirming shift occurred in the same ear and/or the same frequency range (applicable to AAO-HNS). In other words, the original shift could be counted as confirmed not only by a persistent shift in the same ear at the same frequency average but also by a new shift in the other ear at any frequency average. For the OSHA STS TWICE criterion, a true positive was confirmed only by a persistent shift in the same ear on the next audiogram.

The data for classifying true positives from all 15 databases are presented in Table 5–1. The 15-dB TWICE and the 15-dB TWICE 1–4 kHz criteria yielded the two highest percentages of true positive tags—70.9% and 73.3%, respectively. The OSHA STS TWICE criterion yielded 57.0% true positive tags; the remaining criteria yielded between 40.4% and 46.1% true positive tags.

Table 5–1. Classifiable first tags* across 15 databases† and first tags classified as *true positive* for each of the 8-shift criteria‡

Criterion	Number of classifiable first tags	First tags classified as *true positive*	
		Number	%
OSHA STS	958	412	43.0
OSHA STS TWICE	356	203	57.0
AAO-HNS SHIFT	1,291	578	44.8
1972 NIOSH SHIFT	2,268	1,045	46.1
15-dB SHIFT	2,126	858	40.4
15-dB TWICE	1,056	749	70.9
15-dB TWICE 1–4 kHz	726	532	73.3
10-dB AVG. 3–4 kHz	1,175	524	44.6

*Those occurring in comparisons of Tests 2 through 7 back to Test 1.
†N=2,903.
‡Adapted from Royster [1992, 1996].

No criterion evaluated is best in every respect. The relative merits of each are tabulated in Table 5–2. An acceptable criterion should be able to identify promptly a worker with any measurable threshold shift at the most noise-sensitive audiometric frequencies and should tag a reasonably high number of true positives. Relative to the any-frequency criteria, those criteria that average thresholds at two or more audiometric frequencies (i.e., OSHA STS, OSHA STS TWICE, AAO-HNS SHIFT, and 10-dB AVG 3–4 kHz) yield lower numbers of tags with lower percentages of true positives.

For this analysis, the 15-dB TWICE and the 15-dB TWICE 1–4 kHz criteria require that a threshold shift persist on two tests before the worker is identified or "tagged" for meeting the criterion of significant threshold shift; these two criteria result in the two highest percentages of true positives. The 1972 NIOSH SHIFT, which shares with 15-dB TWICE the advantage of not requiring any frequency averaging, uses such a small amount of shift (only 10 dB) at 500 to 3000 Hz that it tags many audiograms that reflect normal testing variability. Thus the 1972 NIOSH SHIFT tags so many workers that it loses its usefulness as a problem identifier. This disadvantage can be partially overcome by increasing the amount of required shift to 15 dB (the 15-dB SHIFT); however, too many workers are still tagged by the 15-dB SHIFT to allow any meaningful followup.

The 15-dB TWICE 1–4 kHz criterion differs from the 15-dB TWICE criterion by excluding shifts at 500 and 6000 Hz. Hearing at the 500-Hz audiometric frequency is

Table 5-2. Advantages and disadvantages of each criterion for significant threshold shift*

Considerations	OSHA STS	OSHA STS TWICE	AAO-HNS SHIFT	1972 NIOSH SHIFT	15-dB SHIFT	15-dB TWICE	15-dB TWICE 1-4 kHz	10-dB AVG. 3-4 kHz
Advantages								
Tags a moderate % of workers			X			X		X
Gives high % of true positive tags						X	X	
Tags workers earliest				X				
No calculation of frequency averages required				X	X	X	X	
Averages noise-susceptible frequencies separately or examines each frequency separately			X	X	X	X		X
Includes all noise-susceptible frequencies			X	X	X	X		
Disadvantages								
Tags a low % of workers	X						X	
Tags such a high % of workers that followup would be impractical				X	X			
Tags workers early in fewer cases		X				X	X	

*Adapted from Royster [1992, 1996].

(Continued)

Table 5–2 (Continued). Advantages and disadvantages of each criterion for significant threshold shift

Considerations	Criteria						
	OSHA STS	OSHA STS TWICE	AAO-HNS SHIFT	1972 NIOSH SHIFT	15-dB SHIFT	15-dB TWICE	15-dB TWICE 1–4 kHz	10-dB AVG. 3–4 kHz
	Disadvantages							
Requires calculations of frequency averages	X	X	X					X
Averages low frequencies that are unlikely to be affected by noise exposure			X					
Averages together frequencies that vary in susceptibility to noise	X	X						
Uses a shift magnitude within the range of normal audiometric variability				X				

Chapter 5. Hearing Loss Prevention Programs (HLPPs)

unlikely to be affected by NIHL, but it may be useful as an indicator of excess ambient noise in the audiometric test booth and as an indicator of the presence of medical ear conditions such as conductive ear pathologies. The 6000-Hz audiometric frequency is one of the three high frequencies (3000, 4000, and 6000 Hz) at which hearing is most likely to be affected soonest and to the greatest degree by NIHL. This audiometric frequency is more susceptible than others to measurement variability if there is inconsistent earphone placement.

Excluding the 500- and 6000-Hz frequencies in the 15-dB TWICE 1–4 kHz criterion reduces the number of tags to less than that for ordinary OSHA STS; also, it does not increase the percentage of true positive tags by any practically important amount (2.4%). This indicates that the shifts at 500 Hz and 6000 Hz that meet the 15-dB TWICE criterion are reliable shifts, not spurious ones. Inclusion of the 6000-Hz frequency is desirable from the standpoint of identifying early NIHL. Therefore, the 15-dB TWICE criterion is preferable to the 15-dB TWICE 1-4 kHz criterion because it identifies a higher number of workers and provides a warning of noise-induced shifts at 6000 Hz, a noise-susceptible test frequency.

The ideal significant threshold shift criterion should tag workers with temporary threshold shifts before they develop into permanent hearing impairment. On the basis of the data analyses presented by Royster [1992, 1996], NIOSH now recommends a modified 15-dB TWICE, 500–6000 Hz criterion. NIOSH recommends an immediate retest after reinstruction and repositioning of the earphones if a 15-dB change in threshold is noted at any frequency. Rink [1989] observed the value of two back-to-back tests and reported that performing an immediate retest reduced the proportion of workers meeting the OSHA STS criterion by more than 70%. Thus, if a monitoring audiogram indicates a 15-dB shift or more in either ear at any one of the test frequencies (500, 1000, 2000, 3000, 4000, or 6000 Hz), the worker should be reinstructed, the earphones refitted, and the retest administered. If the retest shows the same results (i.e., a 15-dB shift or more in the same ear and at the same frequency), the 15-dB TWICE criterion for a significant threshold shift has been met, and the worker should be rescheduled for a confirmation test within 30 days. The confirmation audiogram shall be preceded by a 12-hr period with no exposure to workplace or other loud noises. Hearing protectors shall not be substituted in lieu of the required quiet period.

If the immediate retest is not performed, NIOSH recommends that the significant threshold shift be confirmed by a followup test within 30 days of the testing that showed the significant threshold shift. This followup test is called the confirmation test and is preceded by a 12-hr quiet period. If the significant threshold shift is confirmed and later validated by an audiologist or physician, the confirmation audiogram should be the one with which all subsequent audiograms are compared.

To comply with this recommendation and to provide maximum protection for workers and maximum documentation for employers, NIOSH advocates that audiograms be performed on the following occasions:

1. Before employment or before initial assignment into a hearing hazard work area.

2. Annually for any worker whose noise exposure equals or exceeds 85 dBA as an 8-hr TWA (monitoring audiometry). Annual testing may lead to a number of retests if a significant threshold shift occurs. In addition, it may be a good practice to provide audiometry twice per year to workers exposed to more than 100 dBA, because the most susceptible 10% of a population exposed to daily average noise levels of 100 dBA with inadequate hearing protectors could develop significant hearing loss well before the end of 1 year [NIOSH 1996].

3. At the time of reassignment from a job involving hearing hazards.

4. At the termination of employment.

5.5.1.1 Baseline Audiogram

The baseline audiogram should be obtained within 30 days of enrollment in the HLPP [NIOSH 1972]. It shall be preceded by a minimum of 12 hr of unprotected quiet. Data have supported the concept that following a period of noise exposure, the worker should be provided at least as much time for recovery from temporary threshold shifts as the duration of the noise exposure [Johnson et al. 1976]. Use of hearing protectors should not be considered a substitute for an actual 12-hr quiet period. Use of a mobile testing service should not waive these requirements. It is unacceptable to wait up to a year, as permitted by OSHA [29 CFR 1910.95], for a mobile service to conduct a baseline audiogram, because permanent hearing loss can occur within relatively short periods (months or even days in susceptible workers), especially when high levels of noise are involved [ISO 1990]. If a mobile service cannot meet these time constraints, other arrangements should be made to obtain the baseline audiograms before or promptly after employment.

5.5.1.2 Monitoring Audiograms

Monitoring audiometry shall be conducted no less than annually. Unlike baseline audiometry, these annual tests should be scheduled at the end of, or well into, the work shift so that temporary changes in hearing due to insufficient noise controls or inadequate use of hearing protection will be noted. The results should be compared immediately with the baseline audiogram to check for any change in hearing sensitivity. The collection of audiograms for later batch comparison with baseline audiograms in another location is an unacceptable practice because it does not afford the opportunity to conduct retests or to discuss the findings with workers in a timely manner.

5.5.1.3 Retest Audiograms

As good practice, NIOSH suggests that audiometry be repeated immediately after any monitoring audiogram that indicates a threshold shift of 15 dB or more at 500, 1000, 2000, 3000, 4000, or 6000 Hz in either ear. The worker should be reinstructed and the headphones refitted before conducting the retest. Those who employ the retest strategy will find a significant reduction in the number of workers called back for a confirmation audiogram. The reason is that if the retest audiogram does not show the same shift as the monitoring audiogram, the retest audiogram becomes the test of record and there is no need to call the worker back for a confirmation audiogram.

5.5.1.4 Confirmation Audiograms

Audiometry should be conducted again within 30 days of any monitoring or retest audiogram that continues to show a significant threshold shift. A minimum of 12 hr of quiet shall precede the confirmation audiogram to determine whether the shift is a temporary or permanent change in hearing sensitivity (i.e., a temporary or permanent threshold shift). The use of hearing protectors as a substitute for a quiet environment is not acceptable. Confirmation audiograms indicating persistent threshold shifts shall trigger written notification to the worker and a referral to the audiometric manager for review and determination of probable etiology. This review should explore all possible causes in addition to occupational noise, including age-related hearing loss, familial hearing loss, medical history, nonoccupational noise exposure, etc. [Franks et al. 1989; Stepkin 1993]. Workers showing a threshold shift with a cause other than noise should be counseled by the audiometric manager and referred to their physicians for evaluation and treatment. Workers should also be referred if they meet any of the otologic or medical criteria recommended by AAO-HNS [1983]. Appropriate action should be triggered for workers showing a threshold shift that is determined by the audiometric manager to have occupational noise exposure as the probable cause. Actions shall, at a minimum, include reinstruction and refitting of hearing protectors, additional training in worker responsibilities for effective hearing loss prevention, and/or reassignment to quieter work areas. The audiometric manager should be responsible for making whatever recommendations he or she deems necessary and for seeing that they are carried out.

5.5.1.5 Exit Audiogram

Audiometry should be conducted when a worker leaves employment or is permanently rotated out of an occupational noise exposure at or above 85 dBA as an 8-hr TWA. This exit audiogram, like the baseline, should be performed after a minimum of 12 hr of quiet. The use of hearing protectors as a substitute for quiet is not acceptable.

NIOSH suggests that hearing tests be offered as a health benefit to workers who are not exposed to hazardous noise levels. The tests in these workers can be conducted early in the day—when it is not recommended that noise-exposed employees be tested for changes in hearing thresholds. In addition to providing a valuable internal control group for comparison to the noise-exposed workers, this policy elevates the perceived importance of the HLPP for management and workers [NIOSH 1996].

5.5.2 Audiometers

Audiometers shall, at a minimum, conform to the specifications of the appropriate ANSI standard for Type 4 audiometers [ANSI 1996b], with the additional stipulation that they have the capacity for testing at 8000 Hz. Type 5 audiometers, which only test to 70 dB HTL, are unacceptable for threshold testing within an occupational HLPP.

Audiometers must be kept in calibration for the audiograms to have any value. An audiometer shall receive a functional check (sometimes called a biologic check) each day the instrument is used [Morrill 1986; NIOSH 1996]. This type of calibration check involves obtaining an audiogram from a person with known, stable thresholds and verifying that no changes in HTL exceeding 10 dB have occurred. A bioacoustic simulator check may be substituted for this procedure. In addition, the audiometer attenuator and frequency selection dials should be cycled through while carefully listening for any extraneous noise or distortion that might interfere with testing. The earphone cords should be manipulated to check for any unwanted static or noise. A check for unwanted sounds, such as the presence of the test signal in the nontest earphone, should be made in accordance with section 5.4.2 of ANSI S3.6–1996 *American National Standard Specification for Audiometers* [ANSI 1996b].

An acoustic calibration check shall be performed whenever the functional check indicates a threshold difference exceeding 10 dB in either earphone at any frequency. An acoustic calibration includes checks of output levels, attenuator linearity, and frequency. If the sound pressure levels differ by more than the allowable variances specified by ANSI S3.6–1996 [ANSI 1996b] (or its successor), or if the attenuator linearity differs by more than 1 dB, or if frequency drift exceeds 3%, an exhaustive calibration is necessary [Morrill 1986].

An exhaustive calibration check should be conducted annually or whenever an acoustic calibration indicates the need for such. An exhaustive calibration includes adjusting the audiometer so that it is in compliance with all specifications of ANSI S3.6–1996 [ANSI 1996b] (or its successor) and must be done by an audiometer service technician. It is best to have exhaustive calibrations performed onsite. If the audiometer must be shipped out for this service, an acoustic calibration shall be conducted upon its return to ensure that calibration changes did not occur during shipping [Morrill 1986].

The audiometric test area shall conform to the ambient noise requirements of ANSI S3.1–1991 [ANSI 1991b]. For permanent, onsite test areas, ambient noise levels shall be checked at least annually. For mobile test areas, ambient noise levels should be checked daily or at each new site, whichever is more frequent. Ambient noise levels should be checked with a calibrated sound level meter placed in the test environment at the approximate position that the worker's head will occupy during the test procedure. Some bioacoustic simulators have the capability of measuring ambient noise levels; this is acceptable provided that the unit is placed near the area of the worker's head. All audiometric test equipment as well as lights, heaters, air conditioners, etc. shall be set as they would be during actual testing. The ambient noise levels shall also be measured during audiometric testing; they should be recorded in a log through which they can be traced for each audiogram obtained.

5.6 Use of Hearing Protectors (Component 5)

NIOSH [1996] defines a hearing protector as "anything that can be worn to reduce the level of sound entering the ear." Hearing protectors are discussed more fully in Chapter 6; however, a few brief points should be made here. Hearing protectors are subject to many problems and should be considered the last resort against hazardous noise. Berger [1980] identified several reasons why hearing protectors can fail to provide adequate protection in real-world situations: discomfort, incorrect use with other safety equipment, dislodging, deterioration, and abuse. In addition, hearing protectors generally provide greatest protection from high frequency noise and significantly less protection from low-frequency noise [Berger 1986]. Nevertheless, hearing protectors *can* work as a short-term solution to prevent NIHL if their use is carefully planned, evaluated, and supervised [Berger 1986; Royster and Royster 1990; NIOSH 1996; Franks and Berger, in press].

5.7 Education and Motivation (Component 6)

On November 21, 1983, OSHA promulgated an occupational safety and health standard entitled "Hazard Communication" [29 CFR 1910.1200]. Under the provisions of this standard, employers in the manufacturing sector must establish a comprehensive hazard communication program that includes, at a minimum, container labeling, material safety data sheets, and a worker training program. The hazard communication program is to be written and made available to workers and their designated representatives. Although the Hazard Communication standard does not specifically address occupational noise exposure, the intent of the standard to inform workers of health hazards should apply. Annual training shall be provided to employees exposed to noise levels at or above 85 dBA as an 8-hr TWA. Workers must be informed of the possible consequences of noise exposure and of the various control methods available to protect their hearing. When an HLPP is implemented, workers should be informed of the provisions of the program and the benefits of their full participation in the program.

The success of an HLPP depends largely on effective worker education regarding all aspects of the program. In his review of the hearing conservation literature, Berger [1981] suggests several keys to a successful program: support from management, enforcement of safety policies, education and motivation of the workers, and comfortable and effective hearing protectors. All of these issues depend to some degree on a well-constructed, thorough program of educating and training everyone who is involved in the HLPP.

Obviously, the primary focus of the training component of the HLPP is on the workers. Workers need to be informed about the reasons for and the requirements of the HLPP at the time that they are enrolled. The education process should be ongoing and highlighted by periodic programs focusing on one or more particular aspects of the program. Furthermore, to be optimally effective, education should be tailored to the specific exposure and prevention needs of each worker or group of workers. Education and training will be easily dismissed unless it can be related to each worker's day-to-day functions

[Berger 1981]. Worker education should cover all relevant aspects of the hearing conservation program. At a minimum, the following topics should be included [AOMA 1987; Royster and Royster 1990; NIOSH 1996]:

1. *Requirements of and rationale for the occupational noise standard.*

2. *Effects of noise on hearing.* This should cover both the audiometric effects (i.e., how noise effects show up on an audiogram) and the functional effects (i.e., the impact of NIHL on everyday life).

3. *Company policy for the elimination of noise as a hazard, including noise controls already implemented or planned for the future.* This topic is very important and helps ensure that workers do not accidentally interfere with control measures.

4. *Hazardous noise sources at the worksite.* The discussion should include monitoring procedures, noise maps of the work environment, and use of warning signs as they apply at the site for the workers receiving training.

5. *Training in the use of hearing protectors.* This training should include (a) the purpose of hearing protectors, (b) the types of protectors available and the advantages and disadvantages of each, (c) selection, fitting, use, and care of hearing protectors, and (d) methods for solving common problems associated with hearing protector use. *This training must include supervised, hands-on practice in the proper fitting of hearing protectors.*

6. *Audiometry.* Instruction should include a discussion of the role of audiometry in preventing hearing loss, a description of the actual test procedure, and interpretation and implications of test results. It should be stressed that temporary or permanent threshold shifts indicate *failure* of the HLPP. Workers and managers need to know that threshold shifts may often be traced to inadequate protection resulting from ineffective noise controls and inconsistent use of hearing protectors.

7. *Individual responsibilities for preventing hearing loss.* A discussion of common nonoccupational noise sources and suggested ways of controlling these exposures will further increase the effectiveness of an occupational HLPP [Royster and Royster 1990]. In addition, behavioral research has suggested that it is important to encourage workers' feelings of self-efficacy, control, and personal responsibility for safety and health behavior [Schwarzer 1992].

Despite the emphasis on employee training, management also needs to be educated about the need for and elements of the HLPP. Strong management support is critical to an effective HLPP [AOMA 1987]. This support must be more than just implicit approval of company hearing loss prevention policies. It must be an outward, active show of approval and compliance with the established policies. This support must be clearly evident to lower management, foremen, and workers. Management needs to know the basics of the legal and professional requirements for effective hearing loss prevention as well as the administrative requirements for compliance and the liability consequences of noncompliance. Motivation of upper management may be heightened by emphasizing the possible financial benefits of an effective HLPP on workers' compensation costs, improved productivity, and worker retention [Royster and Royster 1990].

In addition to the workers and managers, members of the hearing loss prevention team must be educated about company policy for the program and their role in it. They must receive appropriate training to enable them to fulfill their duties successfully. This training is *especially* important for those who will be responsible for fitting hearing protectors and training workers in their proper use [Royster and Royster 1990]. If a hierarchy of responsibility exists within the program's team, each member should know his or her place in it. Consultants, including physicians or audiologists who conduct followup examinations, should also be well informed about the company's hearing loss prevention policies to help prevent recommendations or decisions that might conflict with established company policy [Royster and Royster 1986].

Choice of educational and motivational strategies is critical to the success of the training phase of the HLPP. The techniques used and the content selected for presentation must be tailored to the particular needs of the audience [Royster and Royster 1990].

For all groups involved, an effective training program requires both episodic and ongoing educational opportunities. The most useful opportunity for episodic training of the workers occurs at the time of each worker's annual monitoring audiogram. During this time, the worker is most interested in his or her hearing status, and recommendations will have the most relevance. Time should be taken immediately after testing to explain the results of the hearing test, its relationship to the worker's baseline audiogram, and its implications for the adequacy of the worker's hearing protector use. Stable hearing should be praised to reinforce the worker's proper use of noise controls and hearing protectors, and hearing shifts should result in a sincere warning about the need for more *consistent* use of appropriate hearing protectors. The worker must be given the opportunity to ask questions about his or her role in the HLPP and should be encouraged to discuss hearing protector difficulties, etc. [Royster and Royster 1986].

Other opportunities for episodic training also exist. Special training sessions or regularly planned safety meetings should address company policies, results of biennial noise exposure monitoring, overviews of the effect of noise on hearing, and related topics. These training sessions should not be limited to showing a film but should be personally presented by an educator who is knowledgeable about hearing conservation and has an interesting presentation style. Group size should be small enough to permit interaction with the speaker and among the workers. Content should be varied and continually updated [Royster and Royster 1986; NIOSH 1996].

Chapter 5. Hearing Loss Prevention Programs (HLPPs)

In addition to these episodic training sessions, an ongoing educational process should be offered. HLPP personnel, especially the program implementor, should visit the workers' jobsites to see how they are doing. They should talk to workers about the program when they meet them in the halls, at lunch, etc. Posters, bulletin boards, informational pamphlets, etc., can be used as a constant reminder of the importance that the company places on hearing conservation. Contests or awards for effective hearing conservation practices can be used to promote safe behavior [Royster and Royster 1986, 1990]; however, incentive programs should be planned and implemented with full worker participation or they may be perceived by the workers as manipulative attempts by management to control worker behavior [Merry 1995].

5.8 Recordkeeping (Component 7)

Recordkeeping involves creating and maintaining documents on each aspect of the HLPP. This documentation is more than just an exercise in paperwork or computer data entry. Recordkeeping provides the only compelling evidence that the HLPP components were properly, consistently, and thoroughly conducted. Program records are often needed many years after they are collected. If it cannot be established that they are valid, the records are useless. Clearly, documentation needs to be viewed as one of the most critical aspects of an HLPP [Gasaway 1985].

HLPP records are medical records and should be treated with the same degree of integrity and confidentiality. The recordkeeping system should be compatible with the company's general safety and health record system. The company should keep copies of all records, even if a contractor collects the data [NIOSH 1996]. In addition, each worker's noise exposure records, audiometric records, hearing protection records, and training participation records should be cross-referenced so that information about one program component can be readily linked with information about all other program components for that worker. Such cross-referencing is critical to building a total hearing history and establishing the probable cause of any hearing loss should a claim ever be filed [Gasaway 1985; NIOSH 1996].

5.8.1 Noise Exposure Records

Noise exposure records need to include the worker's name, identification number, job code, job description, department, and similar related information such as the current noise exposure level, the date of the last exposure assessment, the monitoring method used, and the name of the person who did the monitoring [NIOSH 1996]. The employee's record should also include the previous noise exposure history. It is useful to include both calculated exposure levels and the raw data from which the calculations were made [Royster et al. 1986].

Noise exposure records should be maintained for a minimum of 30 years, the period that OSHA requires employers to keep other industrial hygiene records [29 CFR 1910.20]. However, it may be prudent to keep noise exposure records even longer. Royster et al.

Noise Exposure

[1986] recommend that exposure records be maintained for the length of employment plus 30 years. Employers might also consult their State workers' compensation agencies. Most States have a statute of limitations for filing a claim for occupational hearing loss; however, some States do not [ASHA 1992]. Prudence dictates a check with State regulations to be certain that records are maintained until it is determined that there will be no further use for them [Royster et al. 1986].

5.8.2 Audiometric Records

Audiometric records need to include the worker's name, identification number, sex, date of birth, and a self-reported worker history. The history should include medical information that may have an impact on hearing status, history of past occupational or military noise exposure, and types of nonoccupational noise exposure [Helmkamp et al. 1984; NIOSH 1996]. Occupational exposure to potentially ototoxic chemicals should also be recorded [Rybak 1992]. Morrill recommends a brief "high-risk" history, which can be readily taken by a technician; this history can then be used as a framework for a more detailed history, as necessary, if the worker is ever referred to an audiologist or physician for further evaluation [Morrill 1986]. The more detailed the history, the more accurately the audiometric manager will be able to determine the actual cause of any threshold shifts.

For each audiometric examination, the test date, time, and hours since the worker's last noise exposure shall be recorded. Audiometric thresholds at all required frequencies should be recorded. The audiometer's make, model, and serial number shall be noted, as well as the dates of the last exhaustive calibration, the last acoustic calibration, the last functional check, and the last check of room ambient noise levels. In addition, the identity of the tester and the tester's subjective assessment of test reliability should be recorded [NIOSH 1996].

Any time a significant threshold shift is documented, the cause determined by the audiometric manager should be recorded. Also, all followup actions should be documented [Gasaway 1985].

Audiometric test results and records of causes of any confirmed shifts should be maintained for the duration of employment plus 30 years, which is the OSHA requirement for worker health records [29 CFR 1910.20]. Other supporting records (e.g., calibration records, ambient noise level checks, etc.) should be maintained for at least 5 years. However, bearing in mind that audiometric records are only as valid as documentation indicates, it may be prudent to keep all supporting records for as long as the thresholds themselves are maintained [Gasaway 1985].

5.8.3 Hearing Protection Records

Hearing protection records should include the types of hearing protectors used, including make, model, and size, as relevant. Records should also be maintained to document training received by the workers in the proper fitting and use of protectors and the consistency of compliance with requirements for wearing hearing protectors [NIOSH 1996]. Hearing protection records should be maintained for a minimum of 30 years; however, each worker's history of hearing protector use should be kept with the audiograms that are maintained for the duration of employment plus 30 years.

5.8.4 Education Records

Education records should include date and type of training provided, who conducted the training, and attendance (if training was a group program) [NIOSH 1996]. Each worker's education and training records should also be maintained for the duration of employment plus 30 years.

5.8.5 Other Records

Other necessary records might include documentation of periodic audits, exposure assessments, plans for engineering and administrative controls, and results of overall program evaluations [NIOSH 1996]. These records and any other documentation relevant to the HLPP should be maintained for a minimum of 30 years.

5.9 Evaluation of Program Effectiveness (Component 8)

The effectiveness of an HLPP should be evaluated in terms of the hearing losses prevented for each worker and the overall rate of hearing loss in the population of workers. This evaluation should occur on a continual basis.

5.9.1 Individual Effectiveness

The effectiveness of the HLPP in preserving workers' hearing is best evaluated through audiometric monitoring of each noise-exposed worker. All workers whose time-weighted noise exposure meets or exceeds 85 dBA shall receive audiometric testing at no cost to the worker at the intervals noted previously under audiometric evaluation. Comparison of a current audiogram with the baseline audiogram will permit the audiometric manager to assess the adequacy of the program elements for that particular worker. Thus each audiogram serves as a marker of the effectiveness of the hearing loss prevention effort for that individual worker. Any apparent changes in hearing indicate a possible failure in the program.

5.9.2 Overall Program Effectiveness

To assess the effectiveness of the HLPP from an overall programmatic level, it is necessary to have an evaluation method that can monitor trends in the population of workers enrolled in the program and thus identify program problems before many individual threshold shifts occur. This evaluation has two parts. The first part evaluates the internal integrity of the audiometric data. A draft ANSI standard currently details a method for such an evaluation—Draft ANSI S12.13-1991, *American National Standard Evaluating the Effectiveness of Hearing Conservation Programs* [ANSI 1991c]. This standard is based on an assumption that year-to-year variability in a population's hearing thresholds reflects the adequacy of the audiometric monitoring program. High variability in sequential thresholds indicates inadequate control of audiometric test procedures, audiometric calibration problems, or poor recordkeeping. Low variability in sequential thresholds indicates a well-controlled program producing results that may be relied on for accuracy and reliability.

The second part of the program evaluation involves comparing the rate of threshold shift among noise-exposed workers to that of persons not exposed to occupational noise. To this end, Melnick [1984] evaluated the efficacy of several methods. The first was based on the OSHA estimation that a noise-exposed population in compliance with the current noise regulations would still demonstrate a prevalence of hearing loss (defined as thresholds exceeding 25 dB at the frequencies of 500, 1000, and 2000 Hz) up to 10% greater than a non-noise-exposed population by the time workers reached retirement (later OSHA calculations have revised this estimate to be 10% to 15%). This method has the obvious disadvantage of delaying evaluation of the HLPP until a number of workers have reached retirement age; by then, however, improvements to the HLPP will be too late to prevent their hearing loss.

Another method involves evaluating the effectiveness of the overall program on the basis of the percentage of workers showing significant threshold shifts. Ideally, this criterion could be based on a control group (i.e., non-noise-exposed) *within the same company*. However, this system requires that *all* workers, whether or not they are noise-exposed, receive annual audiometric evaluations. Others who have investigated the possibility of using the percentage of significant threshold shifts as an evaluation criterion have reported that 3% to 6% [Morrill and Sterrett 1981] or 5% significant threshold shifts [Franks et al. 1989; Simpson et al. 1994] are reasonable incidence rates that can be met by effective programs. Significant threshold shift incidence rates exceeding these percentages might then be considered evidence of a deficient program. One disadvantage of this technique is that it does not account for the effects of other variables (e.g., age, sex, race, and previous noise exposure history) that might affect the significant threshold shift incidence rates if the noise and nonnoise populations differ substantially. Another disadvantage is that this technique does not differentiate possible causes of program deficiencies. Problems could be as likely to be due to poor audiometry as to excessive noise exposure [Melnick 1984; Simpson et al. 1994].

Pell [1972] used an alternative method in evaluating the effectiveness of the hearing conservation program at DuPont. This method involves a longitudinal analysis of the rate of increased hearing loss (10th, 50th, and 90th percentiles) as a function of age for three classes of worker noise exposure: quiet (<85 dBA), low noise (85-94 dBA), and high noise (>94 dBA). Pell [1972] judged his hearing conservation program to be effective by demonstrating that the rate of hearing loss increase with respect to age did not significantly differ among the three noise categories. This system also requires that both noise-exposed and nonexposed workers receive annual audiometric evaluations. Also, because some persons are susceptible to hearing loss at the REL of 85 dBA, it would be preferable to define the quiet group as those exposed to less than 80 dBA.

The U.S. Army Center for Health Promotion and Preventive Medicine (formerly the U.S. Army Environmental Hygiene Agency) evaluates its HLPPs by rating each element and subelement of the program on a five-point scale ranging from maximally compliant to noncompliant. Total points are added across the subelements to achieve a score for that program element; then a total score is computed for the overall program. Well-defined criteria exist for scoring the subelements, but the program evaluator is also given some flexibility in assigning ratings. Such a system is helpful in that it defines strict criteria for every aspect of the program; these must be met to have a fully successful program. However, some of the currently used criteria are not perfect, because the Center has found several highly rated HLPPs to have unacceptably high incidences of significant threshold shifts [Byrne and Monk 1993].

In general, NIOSH suggests that the success of a smaller HLPP be judged by the audiometric results of individual workers. If there is zero tolerance for occupational hearing loss and a commitment to discover the cause of every change in hearing for each person in the HLPP, the overall program effectiveness should be assured. When it is not possible to examine each worker's results to obtain an adequate picture of the program's efficacy (e.g., if records are inaccessible), an overall evaluation criterion is necessary. Currently, no single method is generally accepted for the overall evaluation of HLPPs. Furthermore, no single method stands out as being superior to the rest. Although previous studies have recommended an incidence rate of significant threshold shift of 5% or less as evidence of an effective HLPP [Morrill and Sterrett 1981; Franks et al. 1989; Simpson et al. 1994], NIOSH currently recommends an incidence rate of 3% or less. The 3% rate is calculated by using the data from a population not exposed to occupational noise in Annex C of ANSI S3.44-1996, *American National Standard Determination of Occupational Noise Exposure and Estimation of Noise-Induced Hearing Impairment* [ANSI 1996c]. In the future, it may be preferable to use incidence rates based on the data from the upcoming National Health and Nutrition Examination Survey (NHANES) IV. These data will reflect the hearing of nonoccupational-noise-exposed cohorts that are contemporary to the present workforce enrolled in HLPPs. They will allow consideration of the effects of age, sex, race, and previous exposures to occupational and nonoccupational noises.

5.10 Age Correction

NIOSH does not recommend that age correction be applied to an individual's audiogram for significant threshold shift calculations. Although many people experience some decrease in hearing sensitivity with age, some do not. It is not possible to know who will and who will not have an age-related hearing loss. Thus, applying age corrections to a person's hearing thresholds for calculation of significant threshold shift will overestimate the expected hearing loss for some and underestimate it for others, because the median hearing loss attributable to presbycusis for a given age group will not be generalizable to that experienced by an individual in that age group. The data on age-related hearing losses describe only the statistical distributions in populations. Furthermore, the age-correction tables developed in the 1972 criteria document [NIOSH 1972] (and subsequently included in the 1983 OSHA Hearing Conservation Amendment to the Occupational Noise Standard [48 Fed. Reg. 9738 (1983)]) were based on a cross-sectional study. Longitudinal data were not available, and the age corrections were estimated by calculating trends as a function of the age of each member of the sample. When data from a cross-sectional study are used, the inherent assumption is that a subject who was 20 years old in 1970 can be expected to experience the same age-related hearing loss by the year 2000 that a 50-year-old subject experienced in 1970. This assumption may not be valid because the general health and societal noise exposures of each generation are likely to differ.

The adjustment of audiometric thresholds for aging has become a common practice in workers' compensation litigation. In this application, age corrections reduce the amount of hearing loss attributable to noise exposure, with a consequent reduction in the amount of compensation paid to workers for their hearing losses. However common "age correcting" is and regardless of the extent to which it is applied, it is technically inappropriate to apply population statistics to an individual. Each age correction number is nothing more than a median value from a population distribution. In age-correcting an audiogram, the underlying assumption is that the individual value is given the 50th percentile, when in fact the 10th or 90th percentile may be the correct value. Thus age-correction formulas cannot be applied to determine with certainty how much of an individual's hearing loss is due to age and how much is due to noise exposure.

Age-correcting audiograms obtained as part of an occupational HLPP are even less appropriate. This is not a compensation issue. The purpose of the program is to prevent hearing loss. If an audiogram is age corrected, regardless of the source of the correction values, the time required for a significant threshold shift to be noted will be prolonged. Delaying the identification of a worker with a significant threshold shift is completely contrary to the purpose of an HLPP.

CHAPTER 6
Hearing Protectors

A personal hearing protection device (or hearing protector) is any device designed to reduce the level of sound reaching the eardrum. Earmuffs, earplugs, and ear canal caps (also called semi-inserts) are the main types of hearing protectors. A wide range of hearing protectors exists within each of these categories. For example, earplugs may be subcategorized into foam, user-formable (such as silicon or spun mineral fiber), premolded, and custom-molded earplugs. In addition, some types of helmets (in particular, flight helmets worn in the military) also function as hearing protectors. Refer to Nixon and Berger [1991] for a detailed discussion of the uses, advantages, and disadvantages of each type of protector. Items not specifically designed to serve as hearing protectors (e.g., cigarette filters, cotton, and .38-caliber shells) should not be used in place of hearing protectors. Likewise, devices such as hearing aid earmolds, swim molds, and personal stereo earphones must never be considered as being hearing protective.

Ideally, the most effective way to prevent NIHL is to remove the hazardous noise from the workplace or to remove the worker from the hazardous noise. Hearing protectors should be used when engineering controls and work practices are not feasible for reducing noise exposures to safe levels. In some cases, hearing protectors are an interim solution to noise exposure. In other instances, hearing protectors may be the only feasible means of protecting the worker. When a worker's time-weighted noise exposure exceeds 100 dBA, both earplugs and earmuffs should be worn. It is important to note that using such double protection will add only 5 to 10 dB of attenuation [Nixon and Berger 1991]. Given the real-world performance of hearing protectors [Berger et al. 1996], NIOSH cautions that even double protection is inadequate when TWA exposures exceed 105 dBA.

How much attenuation a hearing protector provides depends on its characteristics and how the worker wears it. The selected hearing protector must be capable of keeping the noise exposure at the ear below 85 dBA. Because a worker may not know how long a given noise exposure will last or what additional noise exposure he or she may incur later in the day, it may be prudent to wear hearing protectors whenever working in hazardous noise. Workers and supervisors should periodically ensure that the hearing protectors are worn correctly, are fitted properly, and are appropriate for the noise in which they are worn [Helmkamp et al. 1984; Gasaway 1985; Berger 1986; Royster and Royster 1990; NIOSH 1996].

Historically, emphasis has been placed on a hearing protector's attenuation characteristics—almost to the exclusion of other qualities necessary for it to be effective. Although those who select hearing protectors should consider the noise in which they will be worn, they must also consider the workers who will be wearing them, the need for compatibility with other safety equipment, and workplace conditions such as temperature, humidity, and atmospheric pressure [Gasaway 1985; Berger 1986]. In addition, a variety of styles should be provided so that workers may select a hearing protector on the basis of comfort, ease of use and handling, and impact on communication [NIOSH 1996; Royster and Royster 1990]. Each worker should receive individual training in the selection, fitting, use, repair, and replacement of the hearing protector [Gasaway 1985; Royster and Royster 1990; NIOSH 1996]. What is the best hearing protector for some workers may not be the best for others [Casali and Park 1990]. The most common excuses reported by workers for not wearing hearing protectors include discomfort, interference with hearing speech and warning signals, and the belief that workers have no control over an inevitable process that culminates in hearing loss [Berger 1980; Helmkamp 1986; Lusk et al. 1993]. Fortunately, none of these reasons present insurmountable barriers. Given adequate education and training, each can be successfully addressed [Lusk et al. 1995; Merry 1996; Stephenson 1996].

Workers and management must recognize the crucial importance of wearing hearing protectors correctly. Intermittent wear will dramatically reduce their effective protection [NIOSH 1996]. For example, a hearing protector that could optimally provide 30 dB of attenuation for an 8-hr exposure would effectively provide only 15 dB if the worker removed the device for a cumulative 30 min during an 8-hr day. *The best hearing protector is the one that the worker will wear.*

Several methods exist for estimating the amount of sound attenuation a hearing protector provides. In the United States, the NRR is required by law [40 CFR 211] to be shown on the label of each hearing protector sold. The NRR was designed to function as a simplified descriptor of the amount of protection provided by a given device. When its use was first proposed, the most typical method used to characterize sound attenuation was the real ear attenuation at threshold (REAT) method, as described in ANSI S3.19–1974 [ANSI 1974]. Sometimes called the octave-band or long method, this method was believed to provide too much information to be useful for labeling purposes; thus a single-number descriptor (NRR) was devised.

The formulas used to calculate the NRR are based on the octave-band, experimenter fit, REAT method. The NRR was intended to be used to calculate the exposure under the hearing protector by subtracting the NRR from the *C-weighted* unprotected noise level. It is important to note that when working with A-weighted noise levels, one must subtract an additional 7 dB from the labeled NRR to obtain an estimate of the A-weighted noise level under the protector. OSHA has prescribed six methods* with which the NRR can be used. (See 29 CFR 1910.95, Appendix B, and descriptions of methods for calculating and using the NRR in *The NIOSH Compendium of Hearing Protection Devices* [NIOSH 1994].)

---

*The OSHA methods are a simplification of NIOSH methods #2 and #3 [NIOSH 1975, 1994; Lempert 1984].

One problem inherent to using single-number descriptors of sound attenuation is the need to ensure that the resulting value does not sacrifice the estimated protection for the sake of simplicity. Thus these calculations will typically *underestimate* laboratory-derived "long methods" for estimating sound attenuation. To get around some of the limitations associated with NRR calculations, other methods have been developed for estimating hearing protector performance. The single-number rating method and the high-middle-low method may be used when a person needs to estimate performance more accurately than possible with the NRR but does not want to resort to octave-band descriptions of sound attenuation. Detailed descriptions of these methods are in *The NIOSH Compendium of Hearing Protection Devices* [NIOSH 1994].

Both NRR and the other hearing protector ratings referred to above are based on data obtained under laboratory conditions in which experimenters fit hearing protectors on trained listeners. As such, these ratings may differ markedly from the noise reduction that a worker would actually experience in the real world. Specifically, studies have repeatedly demonstrated that real-world protection is substantially less than noise attenuation values derived from experimenter-fit, laboratory-based methods. In the late 1970's and early 1980's, two NIOSH field studies found that insert-type hearing protectors in the field provided less than half the noise attenuation measured in the laboratory [Edwards et al. 1979; Lempert and Edwards 1983]. Since the 1970's, additional studies have been conducted on real-world noise attenuation with hearing protectors [Regan 1975; Padilla 1976; Abel et al. 1978; Edwards et al. 1978; Fleming 1980; Crawford and Nozza 1981; Chung et al. 1983; Hachey and Roberts 1983; Royster et al. 1984; Behar 1985; Mendez et al. 1986; Smoorenburg et al. 1986; Edwards and Green 1987; Pekkarinen 1987; Pfeiffer et al. 1989; Hempstock and Hill 1990; Berger and Kieper 1991; Casali and Park 1991; Durkt 1993]. In general, these studies involved testing the hearing thresholds of occluded and unoccluded ears of subjects who wore the hearing protectors for the test in the same manner as on the job. The tests attempted to simulate the actual conditions in which hearing protectors are normally used in the workplace. Table 6–1 compares the NRRs derived from these real-world noise attenuation data with the manufacturers' labeled NRRs or laboratory NRRs. The laboratory NRRs consistently overestimated the real-world NRRs by 140% to 2,000% [Berger et al. 1996]. In general, the data show that earmuffs provide the highest real-world noise attenuation values, followed by foam earplugs; all other insert-type devices provide the least attenuation. From these results, it can also be concluded that ideally, workers should be individually fit-tested for hearing protectors. Currently, several laboratories are exploring feasible methods for this type of fit testing [Michael 1997].

Royster et al. [1996] addressed problems associated with the use of the NRR. These researchers demonstrated that relying on the manufacturer's instructions or the experimenter to fit hearing protectors may be of little value in estimating the protection a worker obtains under conditions of actual use. The Royster et al. [1996] study reported the results of an interlaboratory investigation of methods for assessing hearing protector performance. The results demonstrated that using untrained subjects to fit their hearing protectors provided much better estimates of the hearing protector's noise attenuation in the workplace than using the experimenter to fit them. This method has since been adopted for use by ANSI in ANSI S12.6–1997 [ANSI 1997]. Furthermore, the method has subsequently been endorsed by the NHCA Task Force on Hearing Protector Effectiveness as well as numerous other professional organizations.[†]

OSHA [1983] has instructed its compliance officers to derate the NRR by 50% in enforcing the engineering control provision of the OSHA noise standard. However, NIOSH concurs with the professional organizations cited above and recommends using subject fit data based on ANSI S12.6–1997 [ANSI 1997] to estimate hearing protector noise attenuation. If subject fit data are not available, NIOSH recommends derating hearing protectors by a factor that corresponds to the available real-world data. Specifically, NIOSH recommends that the labeled NRRs be derated as follows:

Earmuffs	Subtract 25% from the manufacturer's labeled NRR
Formable earplugs	Subtract 50% from the manufacturer's labeled NRR
All other earplugs	Subtract 70% from the manufacturer's labeled NRR

For example, measure noise exposure levels in dBC or dBA with a sound level meter or noise dosimeter.

1. When the noise exposure level in dBC is known, the effective A-weighted noise level (ENL) is:

$$ENL = dBC - \text{derated NRR}$$

2. When the noise exposure level in dBA is known, the effective A-weighted noise level is:

$$ENL = dBA - (\text{derated NRR} - 7)$$

To summarize, the best hearing protection for any worker is the removal of hazardous noise from the workplace. Until that happens, the best hearing protector for a worker is the one he or she will wear willingly and consistently. The following factors are extremely important determinants of worker acceptance of hearing protectors and the likelihood that workers will wear them consistently:

- Convenience and availability
- Belief that the device can be worn correctly
- Belief that the device will prevent hearing loss
- Belief that the device will not impair a worker's ability to hear important sounds
- Comfort
- Adequate noise reduction
- Ease of fit
- Compatibility with other personal protective equipment

[†]The following organizations have endorsed the use of the subject fit procedure according to ANSI S12.6: Acoustical Society of America, American Academy of Audiology, American Association of Occupational Health Nurses, American Industrial Hygiene Association (AIHA), American Society of Safety Engineers, ASHA, CAOHC, and NHCA.

Table 6–1. Summary of real-world NRRs achieved by 84% of the wearers of hearing protectors in 20 independent studies*

Type of hearing protector, model, and reference	Test population (number)	Labeled NRR†	NRR84	Weighted mean NRR84‡	Mean NRR84
Foam:					
E-A-R	—	—	—	12.5	13.2
Crawford and Nozza [1981]	58	29	19	—	—
Hachey and Roberts [1983]	31	29	9	—	—
Lempert and Edwards [1983]	56	29	12	—	—
Edwards and Green [1987]	28	29	19	—	—
Edwards and Green [1987]	28	29	14	—	—
Lempert and Edwards [1983]	56	29	5	—	—
Abel et al. [1978]	55	29	9	—	—
Abel et al. [1978]	24	29	9	—	—
Behar [1985]	42	29	14	—	—
Behar [1985]	24	29	16	—	—
Pfeiffer et al. [1989]	69	29	10	—	—
Casali and Park [1991]	10	29	6	—	—
Casali and Park [1991]	10	29	23	—	—
Hempstock and Hill [1990]	72	29	13	—	—
Berger and Kieper [1991]	22	29	20	—	—
Premolded:					
Ultra-Fit	—	—	—	5.8	7.3
Casali and Park [1991]	10	21	4	—	—
Casali and Park [1991]	10	21	17	—	—
Royster et al. [1984]	19	21	5	—	—
Berger and Kieper [1991]	29	21	3	—	—
V–51R	—	—	—	0.1	2.2
Royster et al. [1984]	12	23	3	—	—
Abel et al. [1978]	20	23	2	—	—
Edwards et al. [1978]	84	23	1	—	—
Fleming [1980]	9	23	6	—	—
Padilla [1976]	183	23	-1	—	—

See footnotes at end of table.

(Continued)

Noise Exposure

Table 6-1 (Continued). Summary of real-world NRRs achieved by 84% of the wearers of hearing protectors in 20 independent studies[*]

Type of hearing protector, model, and reference	Test population (number)	Labeled NRR[†]	NRR84	Weighted mean NRR84[‡]	Mean NRR84
Premolded (Continued):					
Accu-Fit or Com-Fit	—	—	—	4.9	4.5
Fleming [1980]	13	26	2	—	—
Abel et al. [1978]	18	26	7	—	—
EP100	—	—	—	2.1	1.5
Crawford and Nozza [1981]	22	26	0	—	—
Edwards et al. [1978]	28	26	-2	—	—
Abel et al. [1978]	45	26	10	—	—
Smoorenburg et al. [1986]	46	26	-2	—	—
NA	—	—	—	1.0	1.0
Regan [1975]	30	NA	1	—	—
Fiberglass:					
Down	—	—	—	3.3	3.5
Lempert and Edwards [1983]	28	15	4	—	—
Edwards et al. [1978]	56	15	3	—	—
POP	—	—	—	7.7	7.8
Lempert and Edwards [1983]	28	22	4	—	—
Behar [1985]	28	22	10	—	—
Pfeiffer et al. [1989]	51	22	7	—	—
Regan [1975]	30	22	10	—	—
Hempstock and Hill [1990]	39	22	8	—	—
Soft	—	—	—	3.4	4.7
Hachey and Roberts [1983]	36	26	1	—	—
Pfeiffer et al. [1989]	12	26	9	—	—
Hempstock and Hill [1990]	32	26	4	—	—
Custom	—	—	—	6.5	5.4
Adcosil:					
Hachey and Roberts [1983]	44	24	4	—	—
NA:					
Crawford and Nozza [1981]	7	NA	7	—	—
Prictear/vent:					
Lempert and Edwards [1983]	56	11	8	—	—
Peacekeeper:					
Lempert and Edwards [1983]	56	15	4	—	—

See footnotes at end of table. (Continued)

Table 6–1 (Continued). Summary of real-world NRRs achieved by 84% of the wearers of hearing protectors in 20 independent studies*

Type of hearing protector, model, and reference	Test population (number)	Labeled NRR†	NRR84	Weighted mean NRR84‡	Mean NRR84
Custom (Continued):					
NA:					
Abel et al. [1978]	48	NA	3	—	—
Regan [1975]	6	NA	4	—	—
Padilla [1976]	230	NA	8	—	—
Semiaural:					
Sound-Ban	—	—	—	9.6	9.3
Behar [1985]	32	17	10	—	—
Casali and Park [1991]	10	19	6	—	—
Casali and Park [1991]	10	19	12	—	—
Earmuffs	—	—	—	13.8	13.8
Bilsom UF–1:					
Hachey and Roberts [1983]	31	25	13	—	—
Casali and Park [1991]	10	25	16	—	—
Casali and Park [1991]	10	25	20	—	—
MSA Mark IV:					
Abel et al. [1978]	47	23	11	—	—
Durkt [1993]	15	23	4	—	—
Optac 4000:					
Pfeiffer et al. [1989]	33	NA	14	—	—
Peltor H9A:					
Pfeiffer et al. [1989]	34	22	14	—	—
Rcal Auralguard III:					
Hempstock and Hill [1990]	42	NA	19	—	—
Norseg:					
Regan [1975]	30	NA	8	—	—
AO 1720:					
Durkt [1993]	11	21	6	—	—
Bilsom 2450:					
Pfeiffer et al. [1989]	11	NA	13	—	—
Clark E805:					
Abel et al. [1978]	17	23	15	—	—
Glendale 900:					
Durkt [1993]	10	21	10	—	—
Optac 4000S:					
Pfeiffer et al. [1989]	10	NA	14	—	—

See footnotes at end of table.

(Continued)

Table 6-1 (Continued). Summary of real-world NRRs achieved by 84% of the wearers of hearing protectors in 20 independent studies[*]

Type of hearing protector, model, and reference	Test population (number)	Labeled NRR[†]	NRR84	Weighted mean NRR84[‡]	Mean NRR84
Earmuffs (Continued):					
Safety 208:					
Abel et al. [1978]	15	22	12	—	—
Safety 204:					
Behar [1985]	9	21	22	—	—
Welsh 4530:					
Regan [1975]	5	25	20	—	—
Miscellaneous:					
Pekkarinen [1987]	71	NA	13	—	—
Safir E/ISF:					
Hempstock and Hill [1990]	20	NA	14	—	—
Miscellaneous:					
Chung et al. [1983]	64	24	18	—	—
Cap Muffs	—	—	—	14.3	14.8
Bilsom 2313:					
Hempstock and Hill [1990]	37	23	16	—	—
Hellberg No Noise:					
Abel et al. [1978]	58	23	11	—	—
Peltor H7P3E:					
Behar [1985]	36	24	13	—	—
AO 1776K:					
Behar [1985]	26	21	14	—	—
Hellberg 26007:					
Hempstock and Hill [1990]	20	NA	18	—	—
Miscellaneous:					
Chung et al. [1983]	37	23	17	—	—
Plug+Muff:					
E-A-R + UF-1:					
Hachey and Roberts [1983]	10	—	25	25.0	25.0

[*]Adapted from Berger et al. [1996].
[†]Abbreviations: NRR = noise reduction rating; NRR84 = NRR achieved by 84% of the wearers of hearing protectors; NA = not available.
[‡]Weighted on the basis of the test population size.

CHAPTER 7

Research Needs

Considerable progress has been made in our understanding of occupational hearing loss prevention. However, additional research is needed to clarify the risks associated with various noise and ototoxic exposures and to reduce the incidence of hearing loss among workers. Furthermore, investigations of possible biological indicators of susceptibility to NIHL would be welcome. For example, although tinnitus is a frequent complaint of the noise-exposed worker, its relationship to permanent hearing loss is not well understood. The additional topics listed in the sections below do not include all areas that would benefit from further investigations, but they represent persistent problems or emerging trends.

7.1 Noise Control

Research is needed to reduce noise exposures through engineering controls in workplaces where the noise exposures are still being controlled primarily by hearing protectors. An HLPP is complex and difficult to manage effectively, and the need for one can be obviated by noise control procedures that reduce noise levels to less than 85 dBA. As important as such noise reduction technologies are, it is equally important to apply traditional noise control engineering concepts to the building of new facilities and equipment. Research also is needed to improve the retrofitting of noise controls to existing operations. A database of effective solutions (best practices) should be created and made accessible to the public.

7.2 Impulsive Noise

Research is needed to define the hazardous parameters of impulsive noise and their interrelationships. These parameters should include amplitude, duration, rise time, number of impulses, repetition rate, and crest factor. In the absence of any other option, impulsive noise is integrated with continuous noise to determine the hazard. Laboratory research with animals and retrospective studies of workers indicate that impulsive noise is more hazardous to hearing than continuous noise of the same spectrum and intensity. However, sufficient data are not available to support the development of damage risk criteria for impulsive noises.

7.3 Nonauditory Effects

Research is needed to define dose-response relationships between noise and nonauditory effects such as hypertension and psychological stress. Studies of hypertension conducted on noise-exposed workers have established a relationship between hypertension and NIHL but have not established a relationship between noise exposure and hypertension. Workplace accidents need to be analyzed to determine whether noise interference with oral communication or audio alarms has been a contributing factor. Technologies must be developed to allow easy identification of warning signals and efficient communication in noisy environments while providing effective hearing protection.

7.4 Auditory Effects of Ototoxic Chemical Exposures

The ototoxic properties of industrial chemicals and their interaction with noise have been investigated for only a few substances. Research in animals is needed to investigate the range of chemicals known to be ototoxic or neurotoxic and to appraise the risk of hearing loss from exposures to these chemicals alone or in combination with noise. Research is needed to support damage risk criteria for combined exposure.

7.5 Exposure Monitoring

NIOSH has been a pioneer in developing an exposure monitoring strategy for air contaminants based on the application of statistical methods [NIOSH 1977]. However, the appropriateness of the strategy for occupational noise exposure has not been determined, and not much research has been conducted in this area since 1977. Limited studies have indicated that a different strategy for monitoring occupational noise exposure may be required [Behar and Plenar 1984; Henry 1992]. Worker exposures to noise must be accurately monitored and appropriate control measures must be implemented when necessary. Several individuals and organizations have proposed different approaches to monitoring noise exposures [Behar and Plenar 1984; CSA 1986; Royster et al. 1986; Hawkins et al. 1991; Henry 1992; Simpson and Berninger 1992; Stephenson 1995]. NIOSH acknowledges the contributions of these individuals and organizations to this important subject and encourages continued effort in the development of exposure monitoring strategies applicable to occupational noise exposure. An important component of HearSaf 2000 is being codeveloped by NIOSH, the United Auto Workers-Ford National Joint Committee on Health and Safety, Hawkwa Group, and James, Anderson and Associates: noise monitoring with emphasis on noise exposure characterizations based on the principles of a task-based exposure assessment model (T-BEAM). The T-BEAM approach stresses the identification of all hazards (including noise) that may be associated with a particular work task. This approach may be especially suitable for mobile or itinerant workers. Additional research is needed to compare these monitoring approaches (including T-BEAM) to determine the best technique for a particular type of worker or work environment.

7.6 Hearing Protectors

The noise attenuation of hearing protectors as they are worn in the occupational environment is usually quite different from that realized in the laboratory. The manufacturer's labeled NRRs (which are currently used by OSHA in determining compliance with the PEL when engineering controls are being implemented or are not feasible) usually do not reflect actual experiences. Thus a pressing need exists for a laboratory method to estimate the noise attenuation obtained with hearing protectors worn in the field. Field research is now needed to validate the new laboratory subject-fit method with onsite fit-testing methods. Research should also lead to the development of hearing protectors that eliminate troublesome barriers by providing increased comfort to wearers as well as improved speech intelligibility and audibility of warning signals. In addition, as new technologies such as active-level dependency and active noise reduction are introduced into personal hearing protection, methods must be developed to describe the effectiveness of these methods alone and when built into passive hearing protectors.

7.7 Training and Motivation

Research is needed in using behavioral survey tools as resources for developing training and education programs that address workers' beliefs, attitudes, and intentions about hearing loss prevention. To date, research in training and motivation has focused on materials and their delivery, with the worker considered the passive receptacle. Research is needed to develop materials and programs that more fully involve the worker in the process and give the worker ownership in the HLPP. Additional methods are also needed to improve the training and motivation of workers who must depend on hearing protection.

7.8 Program Evaluation

Several methods for evaluating the effectiveness of an HLPP are discussed in Chapter 5. No single method is generally accepted as being superior to the rest. Further research and development of methods for evaluating the effectiveness of HLPPs are needed, and the method deemed to have the best balance between accuracy and ease of use should be adopted. All existing methods rely on the results of audiometric testing for evaluating effectiveness of the HLPP. Although audiometric data are crucial for managing an HLPP and evaluating the status of each worker, too much time must pass to build a database of audiograms that can support queries about overall program effectiveness. Methods that do not rely on serial audiograms need to be considered for immediate assessment of program effectiveness. Examples of such methods are observed behaviors that predict the success of a program or questionnaire-type surveys that evaluate workers' beliefs and intents (and correlate with actual behaviors).

7.9 Rehabilitation

Noise and hearing conservation regulations fail to deal with the worker who has developed NIHL. This failure affects policies regarding hearing protector use when speech communication is necessary, the use of hearing aids by hearing-impaired workers in noisy areas, and the use of hearing aids with hearing protectors such as earmuffs. Thus the worker with acquired NIHL is often managed as a casualty who is no longer in the HLPP management system.

Management procedures for workers identified with substantial hearing impairment need to be studied. They would include training in listening strategies, speech reading, and optimal utilization of hearing aids. Research also needs to be directed at developing hearing instruments designed to help workers continue to function in noise while protecting hearing and enhancing communication.

Rehabilitation communication strategies need to be studied. Currently, if hearing-loss-prevention service providers were to suggest that noise-exposed workers with NIHL could benefit from amplification, they would be fired. In such a hostile environment, it is very difficult to define, develop, deliver, and evaluate a rehabilitation program.

References

AAO-HNS [1983]. Otologic referral criteria for occupational hearing conservation programs. Washington, DC: American Academy of Otolaryngology—Head and Neck Surgery Foundation, Inc.

Abel SM, Alberti PW, Riko K [1978]. User fitting of hearing protectors: attenuation results. In: Alberti PW, ed. Personal hearing protection in industry. New York: Raven Press, pp. 315–322.

ACGIH [1995]. 1995–1996 threshold limit values (TLVs) for chemical substances and physical agents and biological exposure indices (BEIs). Cincinnati, OH: American Conference of Governmental Industrial Hygienists.

Aniansson G [1974]. Methods for assessing high frequency hearing loss in every-day listening situations. Acta Otolaryngol (Suppl 320):7–50.

ANSI [1969]. American national standard: methods for the calculation of the articulation index. New York: American National Standards Institute, Inc., ANSI S3.5-1969.

ANSI [1974]. American national standard: method for the measurement of real-ear protection of hearing protectors and physical attenuation of ear muffs. New York: American National Standards Institute, Inc., ANSI S3.19-1974; ASA 1-1975.

ANSI [1978]. American national standard: specification for personal noise dosimeters. New York: American National Standards Institute, Inc., ANSI S1.25-1978; ASA 98-1978.

ANSI [1983]. American national standard: specification for sound level meters. New York: American National Standards Institute, Inc., ANSI S1.4-1983; ASA 47-1983.

ANSI [1985]. American national standard: specification for sound level meters, amendment to S1.4-1983. New York: American National Standards Institute, Inc., ANSI S1.4A-1985.

ANSI [1986]. American national standard: specification for octave-band and fractional-octave-band analog and digital filters. New York: American National Standards Institute, Inc., ANSI S1.11-1986; ASA 65-1986.

ANSI [1991a]. American national standard: specification for personal noise dosimeters. New York: American National Standards Institute, Inc., ANSI S1.25-1991; ASA 98-1991.

ANSI [1991b]. American national standard: maximum permissible ambient noise levels for audiometric test rooms. New York: American National Standards Institute, Inc., ANSI S3.1-1991; ASA 99-1991.

ANSI [1991c]. Draft American national standard: evaluating the effectiveness of hearing conservation programs. New York: American National Standards Institute, Inc., Draft ANSI S12.13-1991; ASA 97-1991.

ANSI [1994]. American national standard: acoustical terminology. New York: American National Standards Institute, Inc., ANSI S1.1-1994; ASA 111-1994.

ANSI [1995] American national standard: bioacoustical terminology. New York: American National Standards Institute, Inc., ANSI S3.20-1995; ASA 114-1995.

ANSI [1996a]. American national standard: measurement of occupational noise exposure. New York: American National Standards Institute, Inc., ANSI S12.19-1996.

ANSI [1996b]. American national standard: specification for audiometers. New York: American National Standards Institute, Inc., ANSI S3.6-1996.

ANSI [1996c]. American national standard: determination of occupational noise exposure and estimation of noise-induced hearing impairment. New York: American National Standards Institute, Inc., ANSI S3.44-1996.

ANSI [1997]. American national standard: methods for measuring the real-ear attenuation of hearing protectors. New York: American National Standards Institute, Inc., ANSI S12.6-1997.

AOMA Committee (American Occupational Medical Association's Noise and Hearing Conservation Committee of the Council on Scientific Affairs) [1987]. Guidelines for the conduct of an occupational hearing conservation program. J Occup Med 29(12):981-982.

ASHA Ad Hoc Committee (American Speech-Language-Hearing Association) [1992]. A survey of states' workers' compensation practices for occupational hearing loss. ASHA 34(March, Suppl 8):2-8.

ASHA Task Force (American Speech-Language-Hearing Association Task Force) [1981]. On the definition of hearing handicap. ASHA 23:293-297.

Atherley G, Johnston N [1981]. Audiometry—the ultimate test of success? Ann Occup Hyg 27(4):427-447.

Atherley GRC [1973]. Noise-induced hearing loss: the energy principle for recurrent impact noise and noise exposure close to the recommended limits. Ann Occup Hyg *16*:183–193.

Atherley GRC, Martin AM [1971]. Equivalent-continuous noise level as a measure of injury from impact and impulse noise. Ann Occup Hyg *14*:11–28.

Behar A [1985]. Field evaluation of hearing protectors. Noise Control Eng J *24*(1):13–18.

Behar A, Plenar R [1984]. Noise exposure—sampling strategy and risk assessment. Am Ind Hyg Assoc J *45*(2):105–109.

Belli S, Sani L, Scarficcia G, Sorrentino R [1984]. Arterial hypertension and noise: a cross-sectional study. Am J Ind Med *6*:59–65.

Berger EH [1980]. EARLOG monographs on hearing and hearing protection: hearing protector performance: how they work—and—what goes wrong in the real world. Indianapolis, IN: Cabot Safety Corporation, EARLOG 5.

Berger EH [1981]. EARLOG monographs on hearing and hearing protection: motivating employees to wear hearing protection devices. Indianapolis, IN: Cabot Safety Corporation, EARLOG 7.

Berger EH [1986]. Hearing protection devices. In: Berger EH, Ward WD, Morrill JC, Royster LH, eds. Noise and hearing conservation manual. Akron, OH: American Industrial Hygiene Association, pp. 319–382.

Berger EH, Kieper RW [1991]. Measurement of the real-world attenuation of E-A-R Foam® and UltraFit® brand earplugs on production employees. Indianapolis, IN: Cabot Safety Corporation, E-A-R 91–30/HP.

Berger EH, Franks JR, Lindgren F [1996]. International review of field studies of hearing protector attenuation. In: Axelsson A, Borchgrevink H, Hamernik RP, Hellstrom P, Henderson D, Salvi RJ, eds. Scientific basis of noise-induced hearing loss. New York: Thieme Medical Publishers, Inc., pp. 361–377.

Boettcher FA, Henderson D, Gratton MA, Danielson RW, Byrne CD [1987]. Synergistic interactions of noise and other ototraumatic agents. Ear Hear *8*(4):192–212.

Bohne BA, Pearse MS [1982]. Cochlear damage from daily exposure to low-frequency noise. St. Louis, MO: Washington University Medical School, Department of Otolaryngology. Unpublished.

Bohne BA, Yohman L, Gruner MM [1987]. Cochlear damage following interrupted exposure to high-frequency noise. Hear Res *29*:251–264.

Bohne BA, Zahn SJ, Bozzay DG [1985]. Damage to the cochlea following interrupted exposure to low frequency noise. Ann Otol Rhinol Laryngol 94(2):122–128.

Botsford JH [1967]. Simple method for identifying acceptable noise exposures. J Acous Soc Am 42(4):810–819.

Brogan PA, Anderson RR [1994]. Industrial noise control process. Paper presented at the Annual Meeting of the National Hearing Conservation Association, Atlanta, GA, February 17–19.

Brown JJ, Brummett RE, Fox KE [1980]. Combined effects of noise and kanamycin. Cochlear pathology and pharmacology. Arch Otolaryngol 106:744–750.

Brown JJ, Brummett RE, Meikle MB, Vernon J [1978]. Combined effects of noise and neomycin: cochlear changes in the guinea pig. Acta Otolaryngol 86:394–400.

Burns W [1976]. Noise-induced hearing loss: a stocktaking. In: Stephens SDG, ed. Disorders of auditory function II. New York: Academic Press, pp. 9–27.

Burns W, Robinson DW [1970]. Hearing and noise in industry. London: Her Majesty's Stationery Office.

Byrne C, Henderson D, Saunders S, Powers N, Farzi F [1988]. Interaction of noise and whole body vibration. In: Manninen O, ed. Recent advances in researches on the combined effects of environmental factors. Tampere, Finland: Py-Paino Oy Printing House.

Byrne D, Monk B [1993]. Evaluating a hearing conservation program: a comparison of the USAEHA method and the ANSI S12.13 method [Abstract]. Spectrum 10 (Suppl 1):19.

Casali JG, Park M [1990]. Attenuation performance of four hearing protectors under dynamic movement and different user fitting conditions. Hum Factors 32(1):9–25.

Casali JG, Park M [1991]. Laboratory versus field attenuation of selected hearing protectors. Sound and Vibration 25(10):26–38.

Ceypek T, Kuzniarz JJ, Lipowczan A [1973]. Hearing loss due to impulse noise: a field study. In: Proceedings of the International Congress on Noise as a Public Health Problem, Dubrovnik, Yugoslavia. Washington, DC: U.S. Environmental Protection Agency, EPA Report No. 550/9–73–008, pp. 219–228.

CFR. Code of Federal regulations. Washington, DC: U.S. Government Printing Office, Office of the Federal Register.

Chung DY, Hardie R, Gannon RP [1983]. The performance of circumaural hearing protectors by dosimetry. J Occup Med 25(9):679–682.

Clark WW, Bohne BA [1978]. Animal model for the 4 kHz tonal dip. Ann Otol Rhinol Laryngol 87(Suppl 57)(No. 4, Part 2):1-16.

Clark WW, Bohne BA [1986]. Cochlear damage: audiometric correlates? In: Collins MJ, Glattke T, Harker LA, eds. Sensorineural hearing loss: mechanisms, diagnosis and treatment. Iowa City, IA: University of Iowa Press, pp. 59-82.

Clark WW, Bohne BA, Boettcher FA [1987]. Effect of periodic rest on hearing loss and cochlear damage following exposure to noise. J Acous Soc Am 82(4):1253-1264.

Cohen A [1973]. Extra-auditory effects of occupational noise. Part II. Effects on work performance. Natl Saf News 108(3):68-76.

Cohen A [1976]. The influence of a company hearing conservation program on extra-auditory problems in workers. J Saf Res 8(4):146-162.

Coles RR [1980]. Effects of impulse noise on hearing—introduction. Scand Audiol Suppl 12:11-13.

Coles RR, Rice CG, Martin AM [1973]. Noise-induced hearing loss from impulse noise: present status. In: Proceedings of the International Congress on Noise as a Public Health Problem, Dubrovnik, Yugoslavia. Washington, DC: U.S. Environmental Protection Agency, EPA Report No. 550/9-73-008, pp. 211-217.

Crawford DR, Nozza RJ [1981]. Field performance evaluation of wearer-molded ear inserts. Unpublished paper presented at the American Industrial Hygiene Conference, Portland, OR, May 29.

CSA [1986]. Procedures for the measurement of occupational noise exposure: a national standard of Canada. Toronto, Ontario, Canada: Canadian Standards Association.

Delin C [1984]. Noisy work and hypertension. The Lancet 2:931.

Durkt G Jr. [1993]. Field evaluations of hearing protection devices at surface mining environments. Pittsburgh, PA: U.S. Department of Labor, Mine Safety and Health Administration, IR 1213.

Earshen JJ [1986]. Sound measurement: instrumentation and noise descriptors. In: Berger EH, Ward WD Morrill JC, Royster LH, eds. Noise and hearing conservation manual. Akron, OH: American Industrial Hygiene Association, pp. 38-94.

Edwards RG, Green WW [1987]. Effect of an improved hearing conservation program on earplug performance in the workplace. Noise Control Eng J 28(2):55-90.

Edwards RG, Hauser WP, Moiseev NA, Broderson AB, Green WW [1978]. Effectiveness of earplugs as worn in the workplace. Sound and Vibration 12(1):12-42.

Edwards RG, Hauser WP, Moiseev NA, Broderson AB, Green WW [1979]. A field investigation of noise reduction afforded by insert-type hearing protectors. Cincinnati, OH: U.S. Department of Health, Education, and Welfare, Public Health Service, Center for Disease Control, National Institute for Occupational Safety and Health, DHEW (NIOSH) Publication No. 79-115.

Eldred KM, Gannon WJ, Von Gierke HE [1955]. Criteria for short time exposure of personnel to high intensity jet aircraft noise. Wright-Patterson Air Force Base, OH: U.S. Air Force, WADC Technical Note 55-355.

Embleton TFW [1994]. Upper limits on noise in the workplace. Report by the International Institute of Noise Control Engineering Working Party. Noise/News Int 2(4):230-237.

EPA [1973]. Public health and welfare criteria for noise. Washington, DC: U.S. Environmental Protection Agency, EPA Report No. 550/9-73-002.

EPA [1974]. Information on levels of environmental noise requisite to protect public health and welfare with an adequate margin of safety. Washington, DC: U.S. Environmental Protection Agency, EPA Report No. 550/9-74-004.

EPA [1981]. Noise in America: the extent of the noise problem. Washington, DC: U.S. Environmental Protection Agency, EPA Report No. 550/9-81-101.

Fechter LD, Young JSY, Carlisle L [1988]. Potentiation of noise induced threshold shifts and hair cell loss. Hear Res 34:39-48.

34 Fed. Reg. 790 [1969a]. Bureau of Labor Standards: occupational noise exposure.

34 Fed. Reg. 7948 [1969b]. Bureau of Labor Standards: occupational noise exposure.

39 Fed. Reg. 37773 [1974a]. Occupational Safety and Health Administration: occupational noise exposure; proposed requirements and procedures. (Codified at 29 CFR 1910.)

39 Fed. Reg. 43802 [1974b]. U.S. Environmental Protection Agency: proposed OSHA occupational noise exposure regulation; request for review and report.

40 Fed. Reg. 12336 [1975]. Occupational Safety and Health Administration: occupational noise exposure—review and report requested by EPA.

46 Fed. Reg. 4078 [1981a]. U.S. Department of Labor: occupational noise exposure; hearing conservation amendment; final rule. (Codified at 29 CFR 1910.)

46 Fed. Reg. 4135 [1981b]. U.S. Department of Labor: occupational noise exposure; hearing conservation amendment; final rule. (Codified at 29 CFR 1910.)

48 Fed. Reg. 9738 [1983]. U.S. Department of Labor: occupational noise exposure; hearing conservation amendment; final rule. (Codified at 29 CFR 1910.)

Fleming RM [1980]. A new procedure for field testing of earplugs for occupational noise reduction [Dissertation]. Cambridge, MA: Harvard University, School of Public Health.

Franks JR, Berger EH [in press]. Hearing protection—personal protection—overview and philosophy of personal protection. In: ILO Encyclopaedia of occupational safety and health. Geneva, Switzerland: International Labour Organization.

Franks JR, Morata TC [1996]. Ototoxic effects of chemicals alone or in concert with noise: a review of human studies. In: Axelsson A, Borchgrevink HM, Hamernik RP, Hellström PA, Henderson D, Salvi RJ, eds. Scientific basis of noise-induced hearing loss. New York: Thieme Medical Publishers, Inc.

Franks JR, Davis RR, Krieg EF Jr. [1989]. Analysis of a hearing conservation program data base: factors other than workplace noise. Ear Hear 10(5):273–280.

Gannon RP, Tso SS, Chung DY [1979]. Interaction of kanamycin and noise exposure. J Laryngol Otol 93:341–347.

Gasaway DC [1985]. Documentation: the weak link in audiometric monitoring programs. Occup Health Saf 54(1):28–33.

Glorig A, Ward WD, Nixon J [1961]. Damage risk criteria and noise-induced hearing loss. Arch Otolaryngol 74:413–423.

Guberan E, Fernandez J, Cardinet J, Terrier G [1971]. Hazardous exposure to industrial impact noise. Ann Occup Hyg 14:345–350.

Haag WM Jr. [1988a]. Engineering source controls can reduce worker exposure to noise. Occup Health Saf 57(4):31–33.

Haag WM Jr. [1988b]. Purchasing power. Appl Ind Hyg 3(9):F22–F23.

Hachey GA, Roberts JT [1983]. Real world effectiveness of hearing protection [Abstract]. Philadelphia, PA: American Industrial Hygiene Conference, May 1983.

Hamernik RP, Henderson D [1976]. The potentiation of noise by other ototraumatic agents. In: Henderson D, Hamernik RP, Dosanjh DS, Mills JH, eds. Effects of noise on hearing. New York: Raven Press, pp. 291–307.

Hamernik RP, Henderson D, Coling D, Slepecky N [1980]. The interaction of whole body vibration and impulse noise. J Acous Soc Am 67(3):928–934.

Hamernik RP, Henderson D, Crossley JJ, Salvi RJ [1974]. Interaction of continuous and impulse noise: audiometric and histological effects. J Acous Soc Am 55(1):117–121.

Hamernik RP, Henderson D, Salvi R [1981]. Potential for interaction of low-level impulse and continuous noise. Wright Patterson Air Force Base, OH: U.S. Air Force Aerospace Medical Research Laboratory, Report No. AFAMRL–TR–80–68.

Harris CM, ed. [1991]. Handbook of acoustical measurements and noise control. 3rd ed. New York: McGraw-Hill, Inc.

Hawkins NC, Norwood SK, Rock JC, eds. [1991]. A strategy for occupational exposure assessment. Akron, OH: American Industrial Hygiene Association.

Helmkamp JC, Talbott EO, Margolis H [1984]. Occupational noise exposure and hearing loss characteristics of a blue-collar population. J Occup Med 26(12):885–891.

Helmkamp JS [1986]. Why workers do not use hearing protection. Occ Health Saf 55(10):52.

Hempstock TI, Hill E [1990]. The attenuations of some hearing protectors as used in the workplace. Ann Occup Hyg 34(5):453–470.

Henderson D, Hamernik RP [1986]. Impulse noise: critical review. J Acous Soc Am 80(2):569–584.

Henderson D, Subramaniam M, Gratton, MA, Saunders SS [1991]. Impact noise: the importance of level, duration, and repetition rate. J Acous Soc Am 89(3):1350–1357.

Henry SD [1992]. Characterizing TWA noise exposures using statistical analysis and normality. Unpublished paper presented at the 1992 Hearing Conservation Conference, Lexington, KY, April 1–4.

Hétu R [1979]. Critical analysis of the effectiveness of secondary prevention of occupational hearing loss. J Occup Med 21(4):251–254.

Hétu R [1982]. Temporary threshold shift and the time pattern of noise exposure. Can Acous 10:36–44.

Holmgren G, Johnsson L, Kylin B, Linde O [1971]. Noise and hearing of a population of forest workers. In: Robinson DW, ed. Occupational hearing loss. London: Academic Press.

Humes LE [1984]. Noise-induced hearing loss as influenced by other agents and by some physical characteristics of the individual. J Acous Soc Am 76(5):1318–1329.

INRS (Institut National de Recherche et de Sécurité) [1978]. Etude des risques auditifs auxquels sont soumis les salaries agricoles en exploitations forestières et en scieries. Vandoeuvre, France: Compte rendu d'étude No. 325-B.

Intersociety Committee [1970]. Guidelines for noise exposure control. J Occup Med 12(1):276–281.

ISO (International Organization for Standardization) [1961]. Acoustics—draft proposal for noise rating numbers with respect to conservation of hearing, speech communication, and annoyance. Geneva, Switzerland: Reference No. ISO/TC 43 #219.

ISO (International Organization for Standardization) [1971]. Acoustics—assessment of occupational noise exposure for hearing conservation purposes. 1st ed. Geneva, Switzerland: Reference No. ISO/R 1999 1971(E).

ISO (International Organization for Standardization) [1990]. Acoustics—determination of occupational noise exposure and estimation of noise-induced hearing impairment. 2nd ed. Geneva, Switzerland: Reference No. ISO 1999 1990(E).

Johansson B, Kylin B, Reopstorff S [1973]. Evaluation of the hearing damage risk from intermittent noise according to the ISO recommendations. In: Proceedings of the International Congress on Noise as a Public Health Problem, Dubrovnik, Yugoslavia. Washington, DC: U.S. Environmental Protection Agency, EPA Report No. 550/9-73-008.

Johnson A, Juntunen L, Nylén P, Borg E, Höglund G [1988]. Effect of interaction between noise and toluene on auditory function in the rat. Acta Otolaryngol (Stockh) 105:56–63.

Johnson DL [1973]. Prediction of NIPTS due to continuous noise exposure. Joint EPA/Air Force study. Wright-Patterson Air Force Base, OH: U.S. Air Force Aerospace Research Laboratory, Report No. AMRL-TR-73-91.

Johnson DL, March AH, Harris CM [1991]. Acoustical measurement instruments. In: Harris CM, ed. Handbook of acoustical measurements and noise control. 3rd ed. New York: McGraw-Hill, Inc.

Johnson DL, Nixon CW, Stephenson MR [1976]. Long-duration exposure to intermittent noises. Aviat Space Environ Med 47(9):987–990.

Jonsson A, Hansson L [1977]. Prolonged exposure to a stressful stimulus (noise) as a cause of raised blood pressure in man. The Lancet 1:86–87.

Kryter KD, Ward WD, Miller JD, Eldredge DH [1966]. Hazardous exposure to intermittent and steady-state noise. J Acous Soc Am 39:451–464.

Kuhn GF, Guernsey RM [1983]. Sound pressure distribution about the human head and torso. J Acous Soc Am 73(1):95–105.

Kuzniarz JJ [1973]. Hearing loss and speech intelligibility in noise. In: Proceedings of the International Congress on Noise as a Public Health Problem, Dubrovnik, Yugoslavia. Washington, DC: U.S. Environmental Protection Agency, EPA Report No. 550/9-73-008.

Kuzniarz JJ, Swierczynski Z, Lipowczan A [1976]. Impulse noise induced hearing loss in industry and the energy concept: a field study. In: Proceedings of the 2nd Conference on Disorders of Auditory Function. Southampton. London: Academic Press.

Lees REM, Roberts JH [1979]. Noise-induced hearing loss and blood pressure. CMA J 120(5):1082–1084.

Lempert BL [1984]. Compendium of hearing protection devices. Sound and Vibration 18(5):26–39.

Lempert BL, Edwards RG [1983]. Field investigations of noise reduction afforded by insert-type hearing protectors. Am Ind Hyg Assoc J 44(2):894–902.

Lempert BL, Henderson TL [1973]. Occupational noise and hearing, 1968 to 1972: a NIOSH study. Cincinnati, OH: U.S. Department of Health, Education, and Welfare, Public Health Service, Center for Disease Control, National Institute for Occupational Safety and Health.

Lusk SL, Ronis DL, Baer LM [1995]. A comparison of multiple indicators. Observations, supervisor report, and self-report as measures of workers' hearing protection use. Eval Health Prof 18(1):51–63.

Lusk SL, Ronis DL, Kerr MJ, Atwood JR [1993]. Test of the health promotion model as a causal model of workers' use of hearing protection. Nursing Res 43(3):151–157.

Malchaire JB, Mullier M [1979]. Occupational exposure to noise and hypertension: a retrospective study. Ann Occup Hyg 22:63–66.

Manninen O, Aro S [1979]. Noise-induced hearing loss and blood pressure. Int Arch Occup Environ Health 42:251–256.

Melnick W [1984]. Evaluation of industrial hearing conservation programs: a review and analysis. Am Ind Hyg Assoc J 45(7):459–467.

Mendez AM, Salazar EB, Bontti HG [1986]. Attenuation measurements of hearing protectors in workplace. Argentina: Laboratorio de Acústica y Luminotecnia C.I.C.

Merry CJ [1995]. Instilling a safety culture in the workplace. In: Proceedings of the National Hearing Conservation Association Conference III/XX, Cincinnati, OH, March 22–25.

Merry CJ [1996]. The role of expectancies in workers' compliance with a hearing loss prevention program. In: Proceedings of the National Hearing Conservation Association Meeting, San Francisco, CA, February 22–24.

Michael K [1997]. A field monitoring system for insert-type hearing protectors. Poster presented at the National Hearing Conservation Association Meeting, Orlando, FL, February 20–22.

Moll van Charante AW, Mulder PGH [1990]. Perceptual acuity and the risk of industrial accidents. Am J Epidemiol *131*(4):652–663.

Morata TC, Dunn DE, Kretschmer LW, Lemasters GK, Keith RW [1993]. Effects of occupational exposure to organic solvents and noise on hearing. Scand J Work Environ Health *19*(4):245–254.

Morrill JC [1986]. Hearing measurement. In: Berger EH, Ward WD, Morrill JC, Royster LH, eds. Noise and hearing conservation manual. Akron, OH: American Industrial Hygiene Association.

Morrill JC, Sterrett ML [1981]. Quality controls for audiometric testing. Occup Health Saf *50*(8):26–33.

NHCA [1987]. Occupational hearing conservationist training guidelines. Des Moines, IA: National Hearing Conservation Association, pp.119–122.

Nilsson R, Lidén G, Sandén Å [1977]. Noise exposure and hearing impairment in the shipbuilding industry. Scand Audiol *6*:59–68.

NIOSH [1972]. NIOSH criteria for a recommended standard: occupational exposure to noise. Cincinnati, OH: U.S. Department of Health, Education, and Welfare, Health Services and Mental Health Administration, National Institute for Occupational Safety and Health, DHEW (NIOSH) Publication No. HSM 73–11001.

NIOSH [1973]. The industrial environment—its evaluation and control. Cincinnati, OH: U.S. Department of Health, Education, and Welfare, Public Health Service, Center for Disease Control, National Institute for Occupational Safety and Health, pp. 533–562.

NIOSH [1975]. Compendium of materials for noise control. Cincinnati, OH: U.S. Department of Health, Education, and Welfare, Public Health Service, Center for Disease Control, National Institute for Occupational Safety and Health, DHEW (NIOSH) Publication No. 75–165.

NIOSH [1976]. Survey of hearing loss in the coal mining industry. Cincinnati, OH: U.S. Department of Health, Education, and Welfare, Public Health Service, Center for Disease Control, National Institute for Occupational Safety and Health, DHEW (NIOSH) Publication No. 76-172.

NIOSH [1977]. Occupational exposure sampling strategy manual. Cincinnati, OH: U.S. Department of Health, Education, and Welfare, Public Health Service, Center for Disease Control, National Institute for Occupational Safety and Health, DHEW (NIOSH) Publication No. 77-173.

NIOSH [1982]. Health hazard evaluation report: Newburgh Fire Department, Newburgh, NY. Cincinnati, OH: U.S. Department of Health and Human Services, Public Health Service, Centers for Disease Control, National Institute for Occupational Safety and Health, HETA 81-059-1045.

NIOSH [1988a]. National occupational exposure survey (NOES), field guidelines. Vol. I. Cincinnati, OH: U.S. Department of Health and Human Services, Public Health Service, Centers for Disease Control, National Institute for Occupational Safety and Health, DHHS (NIOSH) Publication No. 88-106.

NIOSH [1988b]. National occupational exposure survey (NOES), analysis of management interview responses. Vol. III. Cincinnati, OH: U.S. Department of Health and Human Services, Public Health Service, Centers for Disease Control, National Institute for Occupational Safety and Health, DHHS (NIOSH) Publication No. 89-103.

NIOSH [1990]. National occupational exposure survey (NOES), sampling methodology. Vol II. Cincinnati, OH: U.S. Department of Health and Human Services, Public Health Service, Centers for Disease Control, National Institute for Occupational Safety and Health, DHHS (NIOSH) Publication No. 89-102.

NIOSH [1994]. The NIOSH compendium of hearing protection devices. Cincinnati, OH: U.S. Department of Health and Human Services, Public Health Service, Centers for Disease Control and Prevention, National Institute for Occupational Safety and Health, DHHS (NIOSH) Publication No. 95-105.

NIOSH [1996]. Preventing occupational hearing loss—a practical guide. Cincinnati, OH: U.S. Department of Health and Human Services, Public Health Service, Centers for Disease Control and Prevention, National Institute for Occupational Safety and Health, DHHS (NIOSH) Publication No. 96-110.

Nixon CW, Berger EH [1991]. Hearing protection devices. In: Harris CM, ed. Handbook of acoustical measurements and noise control. 3rd ed. New York: McGraw-Hill, Inc., pp. 21.1-21.24.

Noweir MH [1984]. Noise exposure as related to productivity, disciplinary actions, absenteeism, and accidents among textile workers. J Saf Res 15(4):163-174.

Öhrström E, Björkman M, Rylander R [1988]. Noise annoyance with regard to neurophysiological sensitivity, subjective noise sensitivity and personality variables. Psychol Med *18*:605–613.

OMB [1987]. Standard industrial classification manual. Washington, DC: Executive Office of the President, Office of Management and Budget.

OSHA [1983]. CPL 2–2.35A–29 CFR 1910.95(b)(1), Guidelines for noise enforcement; Appendix A. Washington DC: U.S. Department of Labor, Occupational Safety and Health Administration, OSHA Directive No. CPL 2–2.35A (December 19, 1983).

Ostergaard PB [1986]. Physics of sound. In: Berger EH, Ward WD, Morrill JC, Royster LH, eds. Noise and hearing conservation manual. Akron, OH: American Industrial Hygiene Association.

Padilla M [1976]. Ear plug performance in industrial field conditions. Sound and Vibration *10*(5):33–36.

Parvizpoor D [1976]. Noise exposure and prevalence of high blood pressure among weavers in Iran. J Occup Med *18*(11):730–731.

Passchier-Vermeer W [1968]. Hearing loss due to exposure to steady-state broadband noise. Delft, Netherlands: Research Institute for Public Health Engineering, Report 35.

Passchier-Vermeer W [1971]. Steady-state and fluctuating noise: its effect on the hearing of people. In: Robinson DW, ed. Occupational hearing loss. New York: Academic Press.

Passchier-Vermeer W [1973]. Noise-induced hearing loss from exposure to intermittent and varying noise. In: Proceedings of the International Congress on Noise as a Public Health Problem, Dubrovnik, Yugoslavia. Washington, DC: U.S. Environmental Protection Agency, EPA Report No. 550/9–73–008.

Pekkarinen J [1987]. Industrial impulse noise, crest factor and the effects of earmuffs. Am Ind Hyg Assoc J *48*(10):861–866.

Pekkarinen J [1989]. Exposure to impulse noise, hearing protection and combined risk factors in the development of sensory neural hearing loss. Kuopio, Finland: University of Kuopio.

Pell S [1972]. An evaluation of a hearing conservation program. Am Ind Hyg Assoc J *33*(1):60–70.

Pfeiffer BH, Kuhn HD, Specht U, Knipfer C [1989]. Sound attenuation by hearing protectors in the real world. Sankt Augustin, Germany: Berufsgenossenschaftliches Institut für Arbeitssicherheit, BIA Report 5/89.

Phaneuf R, Hétu R, Hanley JA [1985]. A Bayesian approach for predicting judged hearing disability. Am J Ind Med 7(4):343–352.

Prince MM, Stayner LT, Smith RJ, Gilbert SJ [1997]. A re-examination of risk estimates from the NIOSH Occupational Noise and Hearing Survey (ONHS). J Acous Soc Am 101(2):950–963.

Pryor G, Dickinson J, Howd RA, Rebert CS [1983]. Transient cognitive deficits and high-frequency hearing loss in weanling rats exposed to toluene. Neurobehav Toxicol Teratol 5:53–57.

Rebert CS, Sorenson SS, Howd RA, Pryor GT [1983]. Toluene-induced hearing loss in rats evidenced by the brainstem auditory-evoked response. Neurobehav Toxicol Teratol 5:59–62.

Regan DE [1975]. Real ear attenuation of personal ear protective devices worn in industry [Thesis]. Kent, OH: Kent State University.

Rink T [1989]. Clinical review of patterns from 300,000 industrial audiograms. Paper presented at the 1989 Industrial Hearing Conservation Conference, Lexington, KY, April 12–14.

Robinson DW [1968]. The relationships between hearing loss and noise exposure. Teddington, United Kingdom: National Physical Laboratory, NPL Aero Report Ac 32.

Royster JD [1992]. Evaluation of different criteria for significant threshold shift in occupational hearing conservation programs. Raleigh, NC: Environmental Noise Consultants, Inc., NTIS No. PB93-159143.

Royster JD [1996]. Evaluation of additional criteria for significant threshold shift in occupational hearing conservation programs. Raleigh, NC: Environmental Noise Consultants, Inc., NTIS No. PB97-104392.

Royster JD, Royster LH [1990]. Hearing conservation programs: practical guidelines for success. Chelsea, MI: Lewis Publishers, pp. 73–75.

Royster JD, Berger EH, Merry CJ, Nixon CW, Franks JR, Behar A, Casali JG, Dixon-Ernst C, Kieper RW, Mozo BT, Ohlin D, Royster LH [1996]. Development of a new standard laboratory protocol for estimating the field attenuation of hearing protection devices. Part I. Research of Working Group 11, Accredited Standards Committee S12, noise. J Acous Soc Am 99(3):1506–1526.

Royster LH, Royster JD [1986]. Education and motivation. In: Berger EH, Morrill JC, Ward WD, Royster LH, eds. Noise and hearing conservation manual. Akron, OH: American Industrial Hygiene Association, pp. 383–416.

Royster LH, Berger EH, Royster JD [1986]. Noise surveys and data analysis. In: Berger EH, Ward WD, Morrill JC, Royster LH, eds. Noise and hearing conservation manual. Akron, OH: American Industrial Hygiene Association.

Royster LH, Royster JD, Gecich TF [1984]. An evaluation of the effectiveness of three hearing protective devices at an industrial facility with a TWA of 107 dB. J Acous Soc Am 76(2):485–497.

Rybak LP [1992]. Hearing: the effects of chemicals. Otolaryngol Head Neck Surg 106:677–686.

Sataloff J, Vassallo L, Menduke H [1969]. Hearing loss from exposure to interrupted noise. Arch Environ Health 18:972–981.

Schmidt JAW, Royster LH, Pearson RG [1980]. Impact of an industrial hearing conservation program on occupational injuries for males and females [Abstract]. J Acous Soc Am 67:S59.

Schwarzer R [1992]. Self-efficacy: thought control of action. Washington, DC: Hemisphere Publishing Corporation.

Shaw EAG [1985]. Occupational noise exposure and noise-induced hearing loss: scientific issues, technical arguments and practical recommendations. APS 707. Report prepared for the Special Advisory Committee on the Ontario Noise Regulation. NRCC/CNRC No. 25051. National Research Council, Ottawa, Ontario, Canada.

Simpson TH, Berninger S [1992]. Comparison of short- and long-term sampling strategies for fractional assessment of noise exposure. Unpublished paper presented at the Hearing Conservation Conference, Cincinnati, OH, April 3.

Simpson TH, Stewart M, Kaltenback JA [1994]. Early indicators of hearing conservation program performance. J Am Acad Audiol 5:300–306.

Singh AP, Rai RM, Bhatia MR, Nayar HS [1982]. Effect of chronic and acute exposure to noise on physiological functions in man. Int Arch Occup Environ Health 50:169–174.

Smoorenburg GF [1990]. Hearing handicap assessment for speech perception using pure tone audiometry. In: Berglund B, Lindvall T, eds. Noise as a public health problem. Vol. 4. Stockholm, Sweden: Swedish Council for Building Research.

Smoorenburg GF, ten Raa BH, Mimpen AM [1986]. Real-world attenuation of hearing protectors. Soesterberg, Netherlands: TNO Institute for Perception.

Starck J, Pekkarinen J [1987]. Industrial impulse noise: crest factor as an additional parameter in exposure measurements. Appl Acous 20:263–274.

Starck J, Pekkarinen J, Pyykkö I [1988]. Impulse noise and hand-arm vibration in relation to sensory neural hearing loss. Scand J Work Environ Health 14:265–271.

Stephenson MR [1995]. Noise exposure characterization via task based analysis. Paper presented at the Hearing Conservation Conference III/XX, Cincinnati, OH, March 22–25.

Stephenson MR [1996]. Empowering the worker to prevent hearing loss: the role of education and training. In: Proceedings of the National Hearing Conservation Association Meeting, San Francisco, CA, February 22–24.

Stephenson MR, Nixon CW, Johnson DL [1980]. Identification of the minimum noise level capable of producing an asymptotic temporary threshold shift. Aviat Space Environ Med 51(4):391–396.

Stepkin R [1993]. Diagnostics in industry: a professional approach to loss prevention. Paper presented at the 19th Annual Meeting of the National Hearing Conservation Association, Albuquerque, NM, February 18–20.

Sulkowski WJ, Kowalska S, Lipowczan A [1983]. Hearing loss in weavers and dropforge hammermen: comparative study on the effects of steady-state and impulse noise. In: Rossi G, ed. Proceedings of the International Congress on Noise as a Public Health Problem. Milan, Italy: Centro Ricerche e Studi Amplifon.

Sulkowski WJ, Lipowczan A [1982]. Impulse noise-induced hearing loss in drop forge operators and the energy concept. Noise Control Eng 18:24–29.

Suter AH [1978]. The ability of mildly hearing-impaired individuals to discriminate speech in noise. Washington, DC: U.S. Environmental Protection Agency, EPA Report No. 550/9-78-100.

Suter AH [1986]. Hearing conservation. In: Berger EH, Morrill JC, Ward WD, Royster LH, eds. Noise and hearing conservation manual. Akron, OH: American Industrial Hygiene Association, pp. 1–18.

Suter AH [1989]. The effects of noise on performance. Aberdeen Proving Ground, MD: U.S. Army Human Engineering Laboratory. Technical Memorandum 3-89.

Suter AH [1992a]. The relationship of the exchange rate to noise-induced hearing loss. Cincinnati, OH: Alice Suter and Associates, NTIS No. PB93-118610.

Suter AH [1992b]. ASHA monographs on communication and job performance in noise: a review. Rockville, MD: American Speech-Language-Hearing Association, Monograph No. 28, pp. 53–78.

Takala J, Varke S, Vaheri E, Sievers K [1977]. Noise and blood pressure. The Lancet *1*:974–975.

Talbott E, Findlay R, Kuller L, Lenkner L, Matthews K, Day R, Ishii EK [1990]. Noise-induced hearing loss: a possible marker for high blood pressure in older noise-exposed populations. J Occup Med *32*(8):685–689.

Talbott E, Helmkamp J, Matthews K, Kuller L, Cottington E, Redmond G [1985]. Occupational noise exposure, noise-induced hearing loss and the epidemiology of high blood pressure. Am J Epidemiol *121*(4):501–514.

Taylor SM [1984]. A path model of aircraft noise annoyance. Sound and Vibration *96*(2):243–260.

Taylor SM, Lempert B, Pelmear P, Hemstock I, Kershaw J [1984]. Noise levels and hearing thresholds in the drop forging industry. J Acous Soc Am *76*(3):807–819.

Thiery L, Meyer-Bisch C [1988]. Hearing loss due to partly impulsive industrial noise exposure at levels between 87 and 90 dB(A). J Acous Soc Am *84*(2):651–659.

U.S. Air Force [1956]. Hazardous noise exposure. Washington, DC: U.S. Air Force, Office of the Surgeon General, AF Regulation 160–3.

U.S. Air Force [1973]. Hazardous noise exposure. Washington, DC: U.S. Air Force, Office of the Surgeon General, AF Regulation 161–35.

U.S. Air Force [1993]. Hazardous noise program. Washington, DC: U.S. Air Force, AFOSH Standard 48–19.

U.S. Army [1994]. Memorandum (Army hearing conservation program policy) of June 24, 1994, from Frederick J. Erdtmann, Deputy Director, Professional Services, Department of the Army, Office of the Surgeon General, Falls Church, VA, for the Surgeon General Distribution List.

USC. United States code. Washington, DC: U.S. Government Printing Office.

Verbeek JHAM, van Dijk FJH, de Vries FF [1987]. Non-auditory effects of noise in industry. IV. A field study on industrial noise and blood pressure. Int Arch Occup Environ Health *59*:51–54.

Voigt P, Godenhielm B, Ostlund E [1980]. Impulse noise—measurement and assessment of the risk of noise induced hearing loss. Scand Audiol Suppl *12*:319–325.

von Gierke HE, Robinson D, Karmy SJ [1981]. Results of the workshop on impulse noise and auditory hazard. Southampton, United Kingdom: University of Southampton, Institute of Sound and Vibration Research, ISVR Memorandum 618.

Ward WD, ed. [1968]. Proposed damage-risk criterion for impulse noise (gunfire) (U). Washington, DC: National Academy of Sciences, National Research Council Committee on Hearing, Bioacoustics, and Biomechanics.

Ward WD [1970]. Temporary threshold shift and damage-risk criteria for intermittent noise exposures. J Acous Soc Am 48:561–574.

Ward WD [1980]. Noise-induced hearing loss: research since 1973. In: Tobias JV, Jansen G, Ward WD, eds. Proceedings of the Third International Congress on Noise as a Public Health Problem. Rockville, MD: American Speech-Language Hearing Assoc, ASHA Report 10.

Ward WD [1986]. Auditory effects of noise. In: Berger EH, Ward WD, Morrill JC, Royster LH, eds. Noise and hearing conservation manual. Akron, OH: American Industrial Hygiene Association.

Ward WD, Nelson DA [1971]. On the equal-energy hypothesis relative to damage-risk criteria in the chinchilla. In: Robinson DW, ed. Occupational hearing loss. London: Academic Press.

Ward WD, Turner CW [1982]. The total energy concept as a unifying approach to the prediction of noise trauma and its application to exposure criteria. In: Hamernik RP, Henderson D, Salvi R, eds. New perspectives on noise-induced hearing loss. New York: Raven Press.

Ward WD, Turner CW, Fabry DA [1982]. Intermittence and the total energy hypothesis. Paper presented at the 104th meeting of the Acoustical Society of America, Rochester NY, November 11.

Ward WD, Turner CW, Fabry DA [1983]. The total-energy and equal-energy principles in the chinchilla. Poster contribution to the Fourth International Congress on Noise as a Public Health Problem, Turin, Italy.

Wilkins PA, Acton WI [1982]. Noise and accidents—a review. Ann Occup Hyg 25:249–260.

Wu TN, Ko YC, Chang PY [1987]. Study of noise exposure and high blood pressure in shipyard workers. Am J Ind Med 12:431–438.

Yeager DM, Marsh AH [1991]. Sound levels and their measurement. In: Harris CM, ed. Handbook of acoustical measurements and noise control. 3rd ed. New York: McGraw-Hill, Inc.

Young SY, Upchurch MB, Kaufman MJ, Fechter LD [1987]. Carbon monoxide exposure potentiates high-frequency auditory threshold shifts induced by noise. Hear Res 26:37–43.

APPENDIX

A Re-Examination of Risk Estimates from the NIOSH Occupational Noise and Hearing Survey (ONHS)[*]

---

[*]Reprinted with permission of the American Institute of Physics from *The Journal of the Acoustical Society of America*, Volume 101, Issue 2, February 1997, pages 950–963. Further reproduction is prohibited without permission of the copyright holder.

A re-examination of risk estimates from the NIOSH Occupational Noise and Hearing Survey (ONHS)

Mary M. Prince, Leslie T. Stayner, Randall J. Smith, and Stephen J. Gilbert
Education and Information Division, National Institute for Occupational Safety and Health, 4676 Columbia Parkway, Cincinnati, Ohio 45226

(Received 13 July 1994; revised 9 April 1996; accepted 20 September 1996)

This paper describes a new analysis of data from the 1968–72 National Institute for Occupational Safety & Health (NIOSH) Occupational Noise and Hearing Survey (ONHS). The population consisted of 1172 (792 noise-exposed and 380 "controls") predominately white male workers from a cross section of industries within the United States. The analysis focused on how risk estimates vary according to various model assumptions, including shape of the dose-response curve and the amount of noise exposure among low-noise exposed workers (or controls). Logistic regression models were used to describe the risk of hearing handicap in relation to age, occupational noise exposure, and duration exposed. Excess risk estimates were generated for several definitions of hearing handicap. Hearing handicap is usually denoted as an average hearing threshold level (HTL) of greater than 25 dB for both ears at selected frequencies. The frequencies included in the biaural averages were (1) the articulation-weighted average over 1–4 kHz, (2) the unweighted average over 0.5, 1, and 2 kHz, and (3) the unweighted average over 1, 2, and 3 kHz. The results show that excess risk estimates for time-weighted average sound levels below 85 dB were sensitive to statistical model form and assumptions regarding the sound level to which the "control" group was exposed. The choice of frequencies used in the hearing handicap definition affected the magnitude of excess risk estimates, which depended on age and duration of exposure. Although data were limited below 85 dB, an age-stratified analysis provided evidence of excess risks at levels ranging from 80 to 84 dB, 85–89 dB, and 90–102 dB. Due to uncertainty in quantifying risks below 85 dB, new data collection efforts should focus on better characterization of dose-response and longitudinal hearing surveys that include workers exposed to 8-hour time-weighted noise levels below 85 dB. Results are compared to excess risk estimates generated using methods given by ANSI S3.44-1996.
[S0001-4966(97)01102-8]

PACS numbers: 43.50.Qp, 43.64.Wn [GAD]

INTRODUCTION

The most common goal for protecting workers from the auditory effects of occupational noise has historically been the preservation of hearing for speech discrimination. With this protection goal in mind, the National Institute for Occupational Safety and Health (NIOSH) defined hearing handicap as a *biaural average* of hearing levels exceeding 25 dB at the audiometric test frequencies of 1, 2, and 3 kHz and 0.5, 1, and 2 kHz (NIOSH, 1972). Here, the term "biaural average" is used to identify the mean value for the left and right ears. Using these definitions, NIOSH (1972) estimated the excess risk of hearing handicap as a function of age, sound levels and duration of occupational noise exposure. Excess risk, also known as percentage risk, is defined as the percentage of individuals with hearing handicap among individuals exposed to daily 8-hour occupational noise exposure after subtracting the percentage of individuals who would typically incur such a handicap due to aging in an unexposed population. For a 40-year lifetime exposure to average daily (8-hour) noise levels of 80, 85, and 90 dB in the workplace, NIOSH (1972) estimated the excess risk to be 3%, 15%, and 29%, respectively for the biaural average over 1, 2, and 3 kHz. [Unless otherwise noted, "dB" implies an A-weighted 8-hour time-weighted average sound level.] Table I compares the NIOSH (1972) excess risk estimates for the biaural average over 0.5, 1, and 2 kHz to those developed by other organizations at approximately the same time.

Since the publication of the 1972 NIOSH Criteria Document, statistical methods for analyzing categorical data outcomes have been improved to assess risk of disease (Breslow and Day, 1980a). The aim of this paper is to reevaluate the models used to generate excess risk estimates from data collected for the NIOSH 1968–72 Occupational Noise and Hearing Survey (ONHS) (Lempert and Henderson, 1973). Using these newer statistical methods, the paper examines the relationship between exposure to noise and risk of noise-induced hearing handicap (NIHH) and highlights areas of uncertainty in estimating risks. These results will be compared to the 1972 NIOSH analysis (NIOSH, 1972) and to the ANSI S3.44 (ANSI, 1996) standard, which adopted the methods developed by the International Standards Organization (ISO 1971, 1990). The data collected in the NIOSH survey are of continuing interest since they were obtained before hearing protection devices were widely used in the U.S. Observations by NIOSH investigators during sound level surveys and management's impressions of their respective plants did not indicate that participating companies had policies *requiring* hearing protection use. Use of protectors, if available at all, were left to the discretion of the workers. No mass use of hearing protectors was noted in any of the

TABLE I. Comparison of excess risk estimates by organization.[a]

Average daily exposure level (dB)	Excess risk estimates (%) Hearing handicap defined as HTLs > 25 dB for the average of 0.5, 1, 2 kHz		
	NIOSH (1972)	ISO 1999 (1971)	EPA[b]
80	3	0	5
85	15	10	12
90	29	21	22.3
95	43	29	not available

[a]These excess risk estimates are for a 40-year lifetime exposure to noise.
[b]From Federal Register, Vol. 39, No. 244, 1974.

companies surveyed (Cohen, personal communications, 1996).

I. RELEVANCE TO COMPARABLE STUDIES OF NOISE-INDUCED HEARING LOSS

Several investigators (Robinson and Sutton, 1975; Royster and Thomas, 1979; NCHS, 1965; Robinson, 1970; Yerg et al., 1978) have examined the relationship of noise-induced permanent threshold shift (NIPTS) and occupational noise exposure. Studies similar to the NIOSH 1968–72 Noise Survey with respect to time period and methods of data collection include Baughn (1973), Passchier-Vermeer (1968) and Burns and Robinson (1970). These studies will be the main focus of our review of the relevant noise and hearing surveys from this period. These studies have been used by ISO 1999 (1971) and ANSI S3.44 (ANSI, 1996) to estimate the risk of NIHH or NIPTS. Table II presents major study characteristics of each of these studies.

As shown in Table II, only the Baughn (1973) study did not screen their workers for otologic abnormalities. These studies report that their populations were restricted to workers with daily constant levels of steady state noise exposure for the entire length of employment. A review of these studies' limitations has been addressed by Ward and Glorig (1975) and Yerg et al. (1975). They include possible contamination of non-steady state noise exposure in the population and small sample sizes for subjects exposed to continuous steady state for daily sound levels below 90 dB. The Passchier-Vermeer report (1968) reviewed published studies and was not specifically designed to address criteria for a noise standard. The NIOSH study (Lempert and Henderson, 1973) was specifically designed to examine risk of noise-

TABLE II. Overview of selected noise and hearing studies used to assess risk of hearing handicap.

Study	Population examined in risk analysis	Exposure characteristics	Screening of subjects
NIOSH ONHS study[a]	1172 predominately white males from a cross section of industries within the U.S. 792 noise-exposed 380 low noise-exposed	Workers exposed to steady state noise for up to 41 years of exposure to daily noise levels from 80–102 dB. Workers exposed to impact or impulse noise were excluded.	Workers were excluded if they had previous noisy jobs, significant firearm exposure (military or recreational), ear disease or other otologic abnormalities, incomplete job histories or unknown noise exposures.
Baughn, 1973	6,835 audiograms on Caucasian male employees from a Midwestern auto parts plant: 1960–65. Stable work force, light turnover. Employees drawn from surrounding farming-industrial community. Age range: 18–68 yrs.	Workers assigned to three 8-hour TWA exposure levels: 78, 86, and 92 dB: $N=852$–78 dB $N=5150$–86 dB $N=833$–92 dB Age used as uniform measure of exposure duration.	No Otological screening of subjects. 2/3 of available tests were excluded due to significant known or unknown exposures.
Passchier-Vermeer, 1968	4557 Caucasian workers from an industrial population in The Netherlands: 4096 males 461 females	Include only workers with constant noise exposure levels for an 8-hour shift for all exposure years considered.	Workers excluded if they had previous noise exposure during other jobs, otologic abnormalities.
Burns & Robinson, 1970	759 noise-exposed workers and 97 non-noise exposed controls from a variety of occupations. Subjects were volunteers. 422 males 337 females	Exposed daily to steady state noise for periods of up to 50 years.	Excluded individuals with existing or previous ear disease or abnormality, exposure to firing weapons, workers whose noise exposure could not be quantified, and those with language difficulty.

[a]Lempert and Henderson, 1972.

induced hearing handicap as a basis for establishing health based occupational standards. The following summary of the study methods is from a NIOSH technical report by Lempert and Henderson (1973).

II. STUDY METHODOLOGY

A. Survey population

In 1968, the U.S. Public Health Service undertook a nationwide study, called the Occupational Noise and Hearing Survey (ONHS). The study was continued and completed by NIOSH in 1972. The aim of the survey was "to characterize noise exposure levels in a variety of industries, to describe the hearing status of workers exposed to such noise conditions, and to establish a relationship between occupational noise exposure and hearing handicap that would be applicable to general industry." Subjects for the study were recruited through notices at industrial hygiene conferences and through the regional offices of the U.S. Public Health Service. All companies interested in participating were considered if certain priority considerations applied. These included (1) existence of a factory or occupational noise conditions having noise levels relevant to developing noise standards and criteria, and (2) a work force with a wide range of years of exposure to such noise levels.

The data collected in the survey included noise measurements, personal background information, medical and otological data and audiometric examinations. Noise level measurements (using Bruel-Kjaer Sound Level Meters) were taken at different areas of each plant and tape recordings were used for laboratory analysis of noise characteristics. A questionnaire was used to obtain information on each worker's job history, military service, hobbies, and medical history pertinent to ear abnormalities and hearing difficulty. An otoscopic inspection of the ears was also made, usually after the completion of the questionnaire. Measurements of hearing levels (using a Rudmose RA-108 audiometer) for pure tone frequencies of 0.5, 1, 2, 3, 4, and 6 kHz in the right and left ears of the workers were conducted in a Rudmose audiometric travel van (model RA-113). Workers from noisy workplaces were always tested at the beginning of their work shift.

For plants with less than 500 employees, the entire work force was tested. For larger plants, a random sample was selected. Individuals from each plant who worked in offices or other quiet work areas were also included in the survey to provide control data.

B. Screened population for analysis

The survey population was "screened" to exclude individuals with prior noise exposure (from occupational and non-occupational sources) and medical or otologic conditions that might affect a person's risk of hearing loss, independent of occupational noise levels at the time of the survey. Criteria for data exclusion included (1) uncertainty in the noise exposure history or validity of audiometric tests and (2) evidence that hearing loss might have been caused by factors other than occupational noise exposure (e.g., military history, other non-occupational noise exposures, head trauma, other audiological/otologic medical conditions). Workers exposed to noise that was not continuous (e.g., discrete impact sounds or noise with highly variable and unpredictable levels) and all maintenance workers were also excluded. Due to the relatively small number of females in the survey population, all analyses were limited to 1172 males (792 noise-exposed and 380 controls).

C. Variable definitions

1. Definition of hearing handicap

The major outcome of interest is hearing handicap, defined as a biaural average hearing threshold level of greater than 25 dB for a selected set of frequencies. In this analysis, the set of frequencies includes (a) 0.5, 1, and 2 kHz, (b) 1, 2, and 3 kHz and (c) 1, 2, 3, and 4 kHz (herein denoted as 1–4 kHz). The 1–4 kHz frequency average was recommended by an American Speech-Language-Hearing Association (ASHA) Task Force (ASHA, 1981), which focused on the need to include frequencies most affected by noise exposure. The ASHA Task Force recommended that percentage formulas should include hearing threshold levels for 1, 2, 3, and 4 kHz, with low and high fences of 25 and 75 dB, representing 0 percent and 100 percent hearing handicap boundaries, respectively (ASHA, 1981). In this analysis, the ASHA recommendation was modified by calculating a weighted average across frequencies rather than an arithmetic average over the test frequencies of 1, 2, 3, and 4 kHz. Weights were assigned according to frequency specific articulation indexes (ANSI, 1969). The articulation index (AI) is a weighted fraction representing (for a given listening situation) the effective proportion of the speech signal that is available (above a masking noise level or hearing threshold) to a listener for conveying speech intelligibility (ANSI, 1969).

Average hearing threshold levels (HTL_{avg}) using the articulation indexes as weights were calculated [Eq. (1)] and then averaged over both ears:

$$HTL_{avg} = \frac{HTL_{1k}W_1 + HTL_{2k}W_2 + HTL_{3k}W_3 + HTL_{4k}W_4}{W_1 + W_2 + W_3 + W_4},$$
(1)

where, $W_1 = 0.24$, $W_2 = 0.38$, $W_3 = 0.34$, and $W_4 = 0.24$ are the weights at 1, 2, 3, and 4 kHz, respectively. This definition will be referred to as the "1–4 kHz AI average" definition of NIHH.

2. Measurement of noise exposure

Daily 8-hour time-weighted average (TWA) noise exposure was estimated for each worker or worker group using (1) area survey samples, (2) interviews with workmen and supervisors to establish typical workday patterns and (3) time-study charts. These charts segmented the workday into a succession of exposures at specific noise levels and for specified durations. Discussions with both management and workmen were necessary to determine changes in noise exposure over the course of many years. Consideration was given to variations in occupational noise conditions due to placement or relocation of machinery and as well as changes in workers' work routine and locations. The reported noise

TABLE III. Covariates considered for inclusion in the analysis of the NIOSH survey.

Variables	Coding conventions
Age at examination	Continuous variable: age in years
	Categorical:[a]
	17–27 years
	28–35 years
	36–45 years
	45–54 years
	>54 years
Duration of noise exposure	Continuous variable: duration in years
	Categorical:[a,b]
	0–1 years
	2–4 years
	5–10 years
	11–20 years
	>20 years
Sound level, L_{NE}, A-weighted 8-hour, time-weighted Average (TWA) sound level—dB, where L_{NE}=average sound levels for exposed workers; L_0=average sound levels for control population	Continuous variable "Centered" at L_0, dB: $(L_{NE}-L_0)$ L_0 was initially fixed to 79 dB but then estimated in models presented in the text.

[a]Categories were the same as NIOSH, 1972.
[b]In the 1972 NIOSH analysis, those exposed to noise for less than 6 months were coded as "0" for duration of exposure. In the current analysis, controls were coded as "0" for duration of exposure and exposed individuals with less than 6 months of exposure were given a value of 0.25.

levels for the study population represent A-weighted eight hour TWA sound levels calculated assuming a 5 dB exchange rate (i.e., 5 dB increase in sound level is exchanged against a factor of 2 in duration within the workday). All levels were measured with sound level meters set to "slow" response. The A-weighted daily noise levels were available on the 792 noise-exposed individuals but not available for the 380 controls. Although sound levels for the control population were not recorded, they were reported to be below 80 dB (Lempert and Henderson, 1973).

3. Other covariates

Other covariates of interest in this paper were age and duration of exposure in years. The risk of hearing handicap was examined in relation to the covariates defined in Table III. For models that included categorical variables for age (reference: 17–27 years) and duration (reference: 0–1 years), four indicator variables were created for different levels of age and duration exposed (Table III). For models that included continuous variables for duration exposed, all controls were reassigned a duration value of zero because it was assumed that duration has no effect on the hearing of the controls. Exposed individuals with less than six months were coded as 0.25 years (midpoint between 0 and 0.5 years).

D. Statistical models

Logistic regression models were used to analyze hearing handicap, defined as the proportion of individuals whose biaural hearing level is greater than 25 dB for averages over selected frequencies. These logistic regression models were fit using the SAS LOGISTIC procedure (SAS Institute, Inc., 1989) and the nonlinear minimization (NLMINB) routine in S-PLUS (Statistical Sciences, Inc., 1993).

Stratified contingency table analyses (Breslow and Day, 1980a) were performed to assess these data for qualitative evidence of hearing handicap due to exposure to noise after controlling for age. The 2×2 contingency tables were stratified by one year age groups and the prevalence of hearing handicap among the three noise-exposed categories of 80–84 dB, 85–89 dB, and 90–102 dB were contrasted to the prevalence among controls. One-sided tests for detecting increased risks were computed using Mantel–Haenszel methods. Further details of this method are found in Breslow and Day (1980a).

The quantitative relationship between hearing handicap and the covariates (defined below) was modeled using logistic regression methods (Breslow and Day, 1980b). These models can be expressed as

$$p = \Pr(Y=1|X) = \frac{e^{F(X;\alpha,\beta,\phi,L_0)}}{1+e^{F(X;\alpha,\beta,\phi,L_0)}}. \qquad (2)$$

where, p=the expected proportion with average hearing level greater than 25 dB (indicated by $Y=1$), given X. ($Y=0$ indicates an average hearing level is less than or equal to 25 dB);

X =a vector of explanatory variables containing information on age, sound level, and duration of exposure;

$$F(X;\alpha;\beta;\phi,L_0) = \alpha + \beta_1(\text{Age}) + [\beta_{2j}(L_{NE}-L_0)^{\phi}], \qquad (3)$$

where

L_{NE}=A-weighted 8-hour TWA sound level for noise-exposed workers in dB;
L_0=parameter for nominal TWA sound level in control population in dB;
ϕ=shape parameter on dB effect;
α=intercept parameter;
β_1=slope coefficient for age effect;
β_{2j}=the slope coefficient for the jth duration of exposure (years) interval, where $j=1,2,3$ represent exposure intervals of 2–4 years, 5–10 years, and >10 years of exposure, respectively.

1. Model development

The first step in the analysis was to fit several hierarchical logistic regression models and compare nested models using likelihood ratio tests (LRTs) to identify which parameters significantly improved the fit to the data (Fienberg, 1987). The fit of the model to the data was evaluated using a likelihood ratio test and examining the log likelihood statistic, G, which is defined by the expression

$$G = -2\Sigma\{Y \log p + (1-Y)\log(1-p)\}, \qquad (4)$$

where the summation is over all individuals in the sample (Breslow and Day, 1980b).

In general, the lower the value of G, the better the fit between the model and the data. Differences in G statistics for nested models may be interpreted as chi-squares (Breslow and Day, 1980b).

To be consistent with the methodology used in the 1972 NIOSH Noise Criteria Document (NIOSH, 1972), the model was initially fit assuming that the sound level for the control population (L_0) was 79 dB and the shape parameter (ϕ) was 1. This was accomplished by first fitting models with main effects only and then adding interaction terms between (a) duration exposed and daily TWA sound level (L); (b) duration exposed and age; and (c) age and sound level. These interaction terms tested whether there should be allowance for differing slopes by levels of other variables. Models with linear main effect of age, duration exposed, and sound levels were fit with an assumption that all control 8-hour TWA sound levels (L_0) were 79 dB. This assumption was made because individual noise exposure data for controls were unavailable but were known to be less than 80 dB (Lempert and Henderson, 1973). Other models with categorical main effects of age and duration were also examined. The final steps of the analysis involved further model refinements that included (1) assuming there is a nondecreasing relationship of prevalence with sound level and duration; (2) refitting functional forms identified by the LRT strategy accordingly; (3) assuming more flexible models for incorporating the effects of sound level by permitting the shape parameter (ϕ) to vary; (4) permitting the control sound level (L_0) to vary from 79 dB; and (5) conducting sensitivity analyses of the impact of critical assumptions.

A final form of the model was fit such that all the parameters (including L_0 and ϕ) were solved for simultaneously. This model form was fit with the following restriction: the control level, L_0, was bounded at 55 dB and 79 dB. For the final model, a two-sided 90 percent confidence interval was calculated for several noise levels using the parametric percentile bootstrap method (Efron and Tibshirani, 1986; Efron, 1982). The same restrictions on L_0 were applied to 1000 bootstrap samples generated to obtain the confidence limits for excess risk. Graphical displays of bootstrap-based confidence limits were smoothed using localized linear regression smoothers in S-PLUS (Statistical Sciences, Inc., 1993).

2. Excess risk estimation

Excess risk for a particular age is defined as the difference between the risk of hearing handicap for the noise-exposed population, given exposure duration, and the exposure sound level, L_{NE} (where $L_{NE} > L_0$), and the risk of hearing handicap among controls. The excess risk associated with exposure to noise evaluated at a given age was estimated from logistic models using the following relationship:

Excess Risk=Pr[Y=1|age, duration, and intensity

of exposure]−Pr[Y=1|age, control].   (5)

Hence, excess risk is assumed to be equivalent to the increase in risk of hearing handicap associated with noise exposure.

3. Sensitivity analyses

Sensitivity analyses were performed to examine how model assumptions may affect the results (i.e., excess risk estimates). Assumptions evaluated in this analysis included (1) the shape of the dose-response relationship; (2) the sound level, L_0, for the control population; and (3) the effect of using different definitions of hearing handicap. The first two issues were addressed during model development, where each assumption was varied while the other remained fixed.

A comparison of how excess risk estimates varied with different definition of hearing handicap was also examined in this analysis. The new definition (1–4 kHz AI average) was compared to definitions previously used by NIOSH (1972)—biaural hearing levels averaged over 1–3 kHz and 0.5–2 kHz. The analyses of different hearing handicap definitions were based on our final model for each definition of hearing handicap: the model in which the control sound level (L_0) and shape parameter (ϕ) were simultaneously estimated.

III. RESULTS

Figure 1 shows the hearing threshold level distributions (10th, 50th, 90th percentiles) for different frequencies by age and sound level categories for exposed and control workers. All hearing thresholds shown are averages over the left and right ears. Data are classified into five age groups and three noise exposure categories (80–87 dB, 88–92 dB, 92–102 dB). The boundaries for the age and sound level categories were selected to provide adequate sample size (i.e., at least 30) in each cell. Sample sizes for the noise-exposed [$n(NE)$] groups are provided for each graph with median exposure duration. The sample sizes for the controls [$n(C)$] are the same within age groups (shown in top panel of each column). The graphs show similar exposure durations within each age cell and increasing trends for median hearing threshold levels with age and sound level. In all cases, control hearing threshold levels are lower than the noise-exposed population. The tendency of median hearing thresholds to increase with increasing age and sound level is also illustrated. The spread of the distribution (given by the 10th and 90th percentiles) is most marked at 3 and 4 kHz.

A scatter plot of the ONHS data showing years of duration versus TWA sound level, L_{NE}, is presented in Figure 2. The vast majority of the data points are at sound levels above 85 dB. Almost 50% of the noise-exposed population had 8-hour TWA sound levels between 85 and 89 dB, while only 27% were exposed below 85 dB. There are also very few data points corresponding to 40 or more years of noise exposure. This lack of data in the low exposure region (80–84 dB) and among workers with long duration of exposure (> 40 years) imposes limitations for quantifying the risks for workers exposed to noise throughout their working lifetime (e.g., 45 years, assuming a worker starts work at 20 years of age and ends at 65 years).

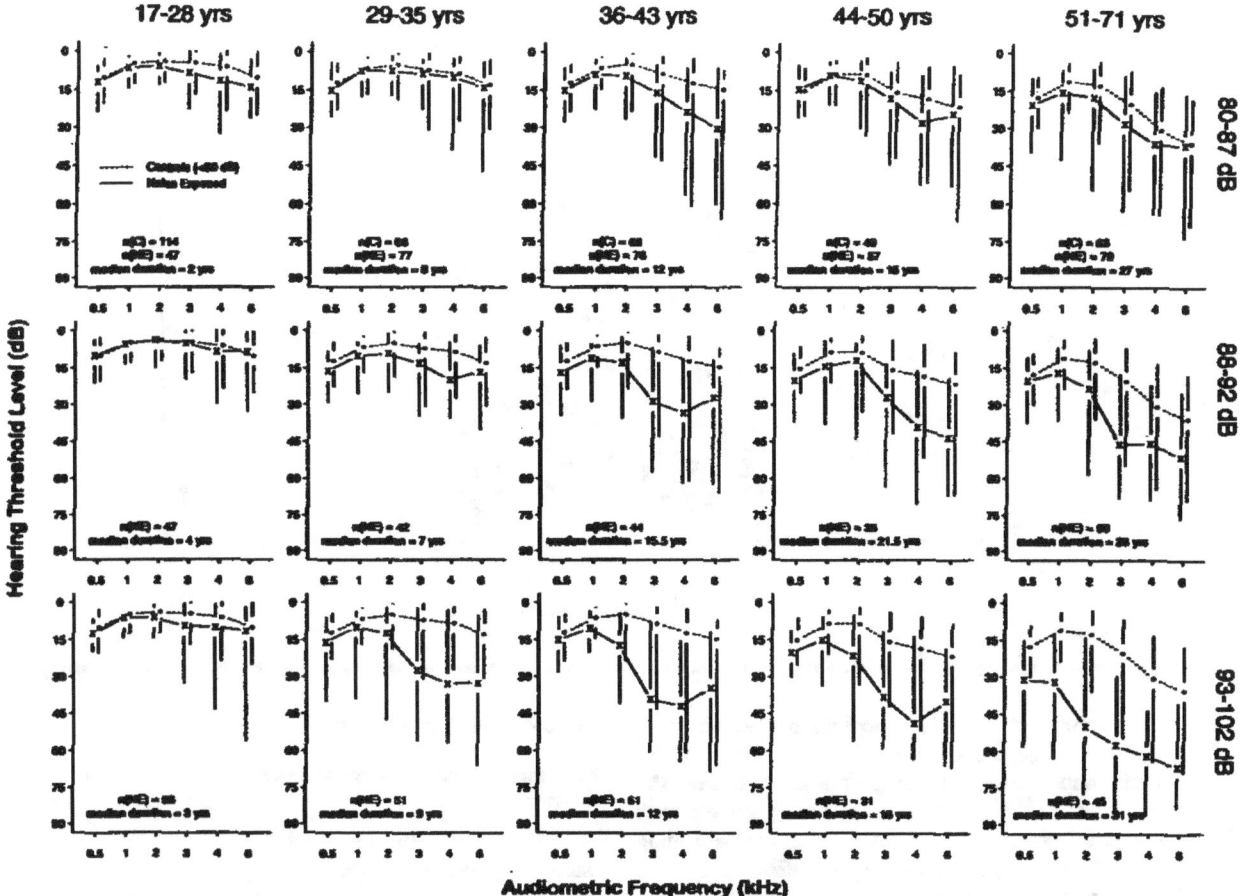

FIG. 1. Distribution of hearing levels (10th, 50th, and 90th percentiles) by age and average daily sound level (L_{NE}) categories from the NIOSH 1968–72 survey.

Despite the limited amount of data in the low exposure region, the Mantel–Haenszel age-stratified analysis provided evidence of positive excess risk associated with sound levels ranging from 80 to 84 dB ($p=0.02$), as well as 85 to 89 dB ($p=0.02$) and 90 to 102 dB ($p<0.001$).

Age was found to be a highly significant predictor of hearing handicap due to noise whether it was modeled using a continuous variable ($\chi^2=211$, $df=1$) or a set of categorical variables ($\chi^2=213$, $df=4$). The fitted categorical effects for age suggested a linear trend (data not shown). This trend was also apparent when models including sound level and duration were fit. Therefore, the simpler models with linear effects for age (as a continuous variable) were subsequently considered in the final models. The addition of either years of exposure or sound level (L_{NE}) significantly improved the fit of the model containing age. The addition of both terms further increased the goodness of fit. A statistically significant interaction ($\chi^2=29.6$, $df=4$) was observed between sound level and categories of years of exposure. No significant interactions between age and duration exposed, nor age and sound level were observed in this data set.

Based on this preliminary analysis, the best fitting linear model is a function of continuous age, categorical levels of duration of exposure, and sound level. However, this model initially appeared to be inappropriate for risk assessment because the excess risk of hearing handicap predicted by this model decreased over limited ranges of sound level and duration of exposure. For example, the parameter estimates of this model suggested that the risk of hearing handicap was lower for individuals with greater than 20 years of exposure than it was for individuals with 11–20 years of exposure when the sound level was above 90 dB. We found no statistically significant difference between the fit of the model that combined the two highest duration categories (11–20 years and >20 years combined to >10 years) and the model with separate parameters for each duration category. This suggested that risks remain essentially flat after 10 years of exposure, and that these two categories could be combined. This initial model was further refined to describe predicted risks of hearing handicap as a nondecreasing function of exposure duration and sound level. The models also assume that the effects of sound level depend on durations greater than or equal to two years.

To test whether a linear sound level effect ($\phi=1$) adequately described the relationship between noise exposure and risk of hearing handicap, higher order terms for the sound level effect were tested in the analysis. Using a quadratic sound level term for exposure ($\phi=2$) appreciably im-

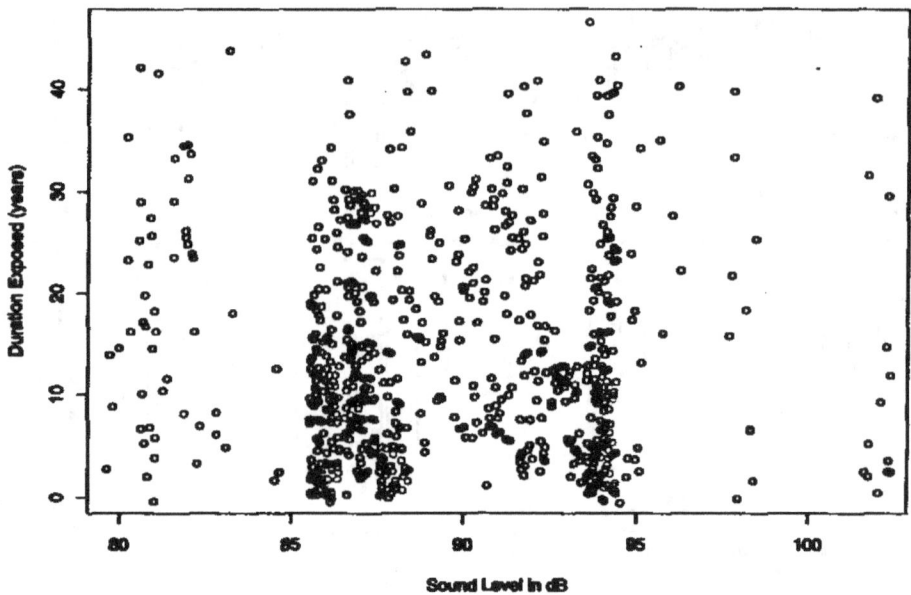

FIG. 2. Scatter plot of exposure sound levels (L_{AE}) versus exposure duration of 792 noise-exposed workers from the NIOSH 1968–72 survey.

proved the goodness of fit of the model relative to the linear model. Using a cubic sound level term ($\phi=3$) resulted in only a slight improvement in the goodness of fit over the quadratic model. The final results from fitting models with linear, quadratic, or cubic sound level terms and assuming control sound levels, L_0, of 79 dB are presented in Table IV. Also shown are the results from fitting a final model in which the control value and the shape parameter were found to be 73 dB and 3.4, respectively (model 4, Table IV). Model 4 is denoted as the "best fitting model," because it produced the best fit to the data. These results indicate considerable variability in excess risk estimates depending on model form and is likely due to lack of data at lower sound levels. This was most marked at average daily sound levels less than or equal to 85 dB. Figure 3 presents excess risk estimates with smoothed 90 percent confidence limits for 65-year-old males with greater than 10 years of exposure as a function of sound level for the "best fitting model" (model 4).

A. Sensitivity analyses

1. Assumption regarding control 8-hour TWA sound levels

To examine the sensitivity of risk estimates to the assumed sound level for the control group, the value of L_0 was varied from 60 to 79 dB and optimum values of the shape parameter, ϕ, were estimated. As L_0 is varied, there is very little variation in the log likelihood statistic, G, whereas the excess risk estimate for noise exposure at a level of 80 dB varies between 0.06 and 2.9 (Table V). The results also show that the optimum value of ϕ decreases considerably as the assumed value of L_0 increases. This analysis suggests that information regarding the distribution of occupational sound levels within the control population is important in estimating the risk of noise-induced handicap in noise-exposed populations. The variability of excess risk estimates below 85 dB seen in Fig. 3 may be attributed to the lack of accurate

TABLE IV. Excess risk percent of noise-induced hearing handicap for workers aged 65 with 10 or more years of noise exposure at various time-weighted average sound levels for linear, quadratic, cubic, and best fitting models.

Exposure sound level (L_{AE}) in dB	Excess risk estimates for various models				
	Quadratic (model 2) $\phi=2$	Cubic (model 3) $\phi=3$	Best fitting (model 4)[a] $\phi=3.4$	Linear dB models ($\phi=1$)	
				Present analysis	NIOSH (1972)
80	0.2	0.02	1.2	3.4	3
85	8.3	3.2	7.6	19.6	15
90	24.5	17.8	22.3	32.2	29
95	28.5	36.2	38.3	40.6	43
100	44.1	41.2	44.0	45.5	56

[a]Risk estimates can be generated using the following equation: $Logit\ [Pr(Y > 25\ dB\ HL)] = -5.0557 + 0.0812(Age) + [\beta_j\ (Duration=1)]*[(L_{AE}-L_0)/(102-73)]^\phi$, where, $\beta_j = 2.6653, 3.989,$ and 6.4206, respectively, for the jth duration of exposure for 2–4 years, 5–10 years, and >10 years, respectively and Y is the AI-weighted binaural average over 1–4 kHz. For the best fitting model, ϕ was estimated to be 3.4 and $L_0 = 73$ dB. The term $(102-73)$ in the denominator of the coefficient describing the effect of duration and sound level, standardizes the exposure term such that the maximum exposure equals one. This was done for ease of comparison to models with differing estimates for L_0 and ϕ.

TABLE V. Excess risk percent of hearing handicap from logistic regression models assuming different sound level values for controls with corresponding shape parameters: Male workers aged 65 with duration exposure greater than 10 years.

Exposure sound level (L_{NE}) in dB	Control sound levels (L_R) in dB and corresponding shape parameters (ϕ)				
	60 (ϕ=5.46)	65 (ϕ=4.67)	70 (ϕ=3.88)	75 (ϕ=3.10)	79 (ϕ=2.49)
80	2.9	2.4	1.8	0.8	0.06
85	9.6	9.1	8.3	7.0	5.2
90	23.4	23.2	22.8	22.1	21.0
95	39.2	39.0	38.6	38.1	37.3
100	45.2	44.9	44.4	43.6	42.6
Log likelihood statistic, G	1039.794	1039.715	1039.645	1039.631	1039.754

sound level data among control subjects and the sparseness of the data for workers exposed at sound levels below 85 dB.

2. Definition of hearing handicap

To examine whether excess risk estimates varied by the definition of hearing handicap used, we compared the 1–4 kHz AI average definition to two other definitions using the same fence (> 25 dB HL), the unweighted biaural frequency averages of 0.5–2 kHz and 1–3 kHz. All three definitions were examined using a model that included age and a dose metric effect defined as $(L_{NE}-L_0)^\phi$ times duration categories (e.g., 2–4, 5–10, and >10 years). The resultant estimated shape parameters for the 0.5–2 kHz and 1–3 kHz biaural averages were 4.5 and 4.9, respectively, with L_0 equal to 55 dB for both.

Under these models, excess risk estimates were affected by both the definition of hearing handicap and the age of the worker. We also found that changing from the articulation index to a simple average of 1–4 kHz did not substantially affect excess risk estimates (results not shown). For workers aged 65 years (with $>$ 10 years of exposure), excess risks for the 1–3 kHz definition are higher than excess risk for the new definition, particularly for sound levels above 85 dB (Fig. 4A). However, among workers aged 45 with similar years of exposure, excess risk estimates are similar for all sound levels for the 1–3 kHz definition and the new definition (Fig. 4B). For younger workers (aged 30 years) with 5 to 10 years of exposure, excess risk estimates for the definitions that included 3 kHz and/or 4 kHz, are similar for all sound levels (Fig. 4C).

IV. DISCUSSION

The results of these analyses indicate that there is an excess risk of noise-induced hearing handicap (NIHH) at 8-hour time-weighted average (TWA) sound levels greater than or equal to 85 dB. The excess risk below 85 dB was not well defined in our analysis. However, the Mantel–Haenszel test result suggests that there is a positive and statistically significant excess risk at levels between 80 and 84 dB.

These findings also indicate two major areas of uncertainty for quantifying the risk of noise-induced hearing handicap. The first concerns the sensitivity of the analysis to the assumed sound level for the control group (L_0). The second relates to the shape of the dose-response relationship between the sound levels among the noise-exposed group (L_{NE}), duration exposed, and the risk of NIHH. Risk estimates were found to vary considerably for values of L_{NE} below 85 dB, depending on the assumed control sound level (L_0), and the shape parameter (ϕ) for the sound level effect (e.g., linear, quadratic, or cubic) in the models.

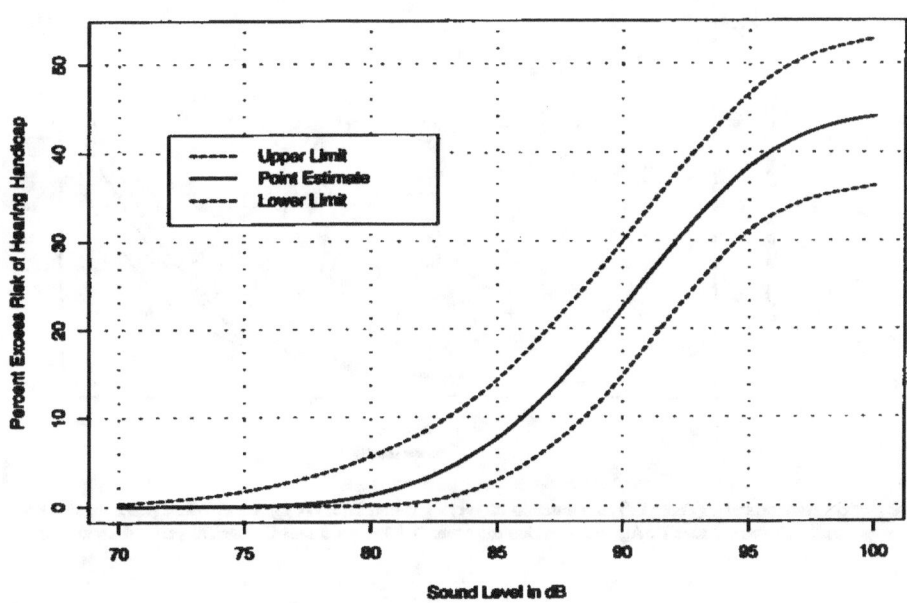

FIG. 3. Excess risk (percent) of hearing handicap (AI-weighting, 1–4 kHz) and bootstrap-based 90% confidence limits from model 4 (Table IV) for 65-year-old males exposed for greater than 10 years to varying levels of noise (L_{NE}).

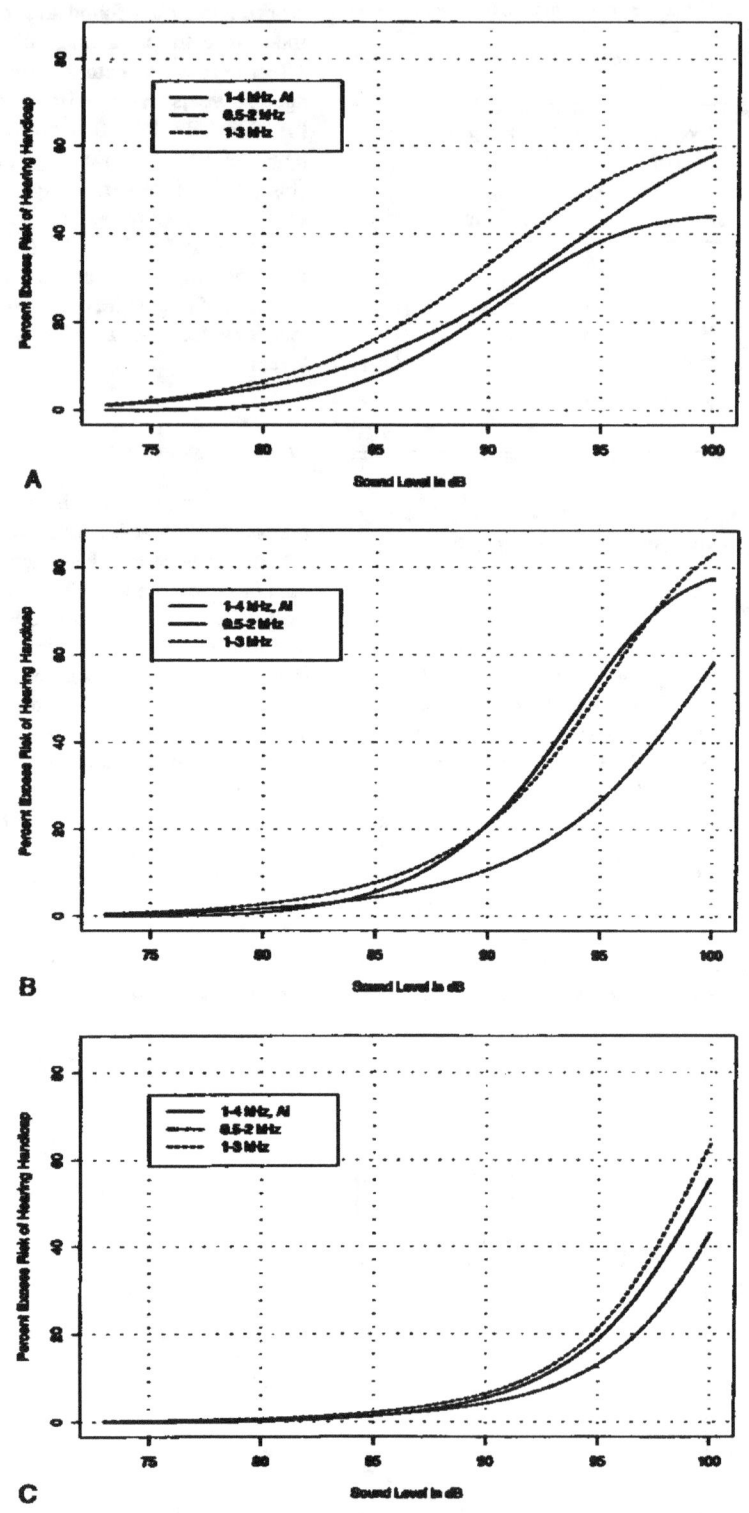

FIG. 4. Excess risk percent from model 4 (Table IV) as a function of varying sound levels (L_{eq}) for different definitions of hearing handicap. Panel A: Age 65 years, duration exposure >10 years. Panel B: Age 45, duration exposure >10 years. Panel C: Age 30 years, duration exposure 5–10 years.

The previous NIOSH (1972) estimate of excess risk for a 40-year working lifetime of exposure to noise was approximately 15 percent at 85 dB. A linear regression model of log hearing levels was used in the previous analysis (NIOSH, 1972) to estimate the risk of hearing handicap. NIHH was defined as an average biaural hearing level greater than 25 dB based on unweighted averages of 0.5–2 kHz or 1–3 kHz. The model described in the 1972 NIOSH criteria document (NIOSH, 1972) is mathematically equivalent to a probit model in which the risk of a hearing level greater than 25 dB is of interest. The results from the previous NIOSH analysis (NIOSH, 1972) also appear to be consistent with the assumption that the control group was exposed to sound levels near 79 dB.

It is clear that models which include a quadratic or cubic effect for the sound level effect fit significantly better than the linear effect model and produce lower excess risk estimates for sound levels below 85 dB than similar models used in the 1972 NIOSH analysis (NIOSH, 1972). As shown in Table IV, the point estimates of excess risk at 85 dB from the quadratic and cubic models are 8 percent and 3 percent, respectively. The quadratic and cubic models fit better than the linear model, mainly due to the effect of sound level in the low exposure region. For sound levels less than or equal to 90 dB, the excess risk estimates from fitting a linear model (Table IV) are slightly higher than those in the NIOSH (1972) analysis. Thus, the disparity in excess risk estimates presented in Table IV may be attributed primarily to the different functional forms (i.e., shape of the sound level effect) of the fitted models. The logistic model used in this analysis assumes the existence of a plateau in risk after 10 years of exposure duration.

The analysis comparing different indicators of NIHH show that patterns of excess risk as a function of average daily sound level depend on age. Differences in excess risk were nominal for the 1–4 kHz average, irrespective of whether HTLs were weighted by the frequency-specific articulation indexes. These differing results by age may be attributable to the fact that the effect of aging on risk of hearing handicap may overshadow any incremental increases in excess risk due to noise exposure. In the upper range of duration and sound level, the dose-response curve shows signs of a plateau effect. The analysis also suggests that the effect of sound intensity and duration of exposure is dependent on frequency. Hearing damage at 3 and 4 kHz is expected to occur sooner than loss at lower frequencies (0.5, 1, or 2 kHz). Definitions that exclude the higher frequencies tend to be less sensitive to noise damage and may require longer durations of exposure to a given sound level to see significant excess risks in the population.

Figure 4A and B suggests that the most suitable definition of hearing handicap may depend on the population characteristics, such as age, exposure duration, and degree of hearing handicap already accrued, as well as whether one chooses to identify preclinical or later stages of hearing handicap. The addition of the most sensitive frequencies to a hearing handicap definition is a valid option if the goal is to have a measure that addresses both prevention and identification of hearing handicap.

A. Data limitations

The cross-sectional design of this study presented limitations for estimating the risk of noise-induced hearing handicap. For example, the 8-hour TWA sound levels, L_{NE}, were determined at one point in time and are assumed to be representative of exposure over the entire length of an employee's job experience. This may have introduced a substantial source of error in the estimation of L_{NE}. As a means of reducing this error, the screened ONHS population included only workers who remained in the same job for the entire time that they worked at the study facility. These workers were then assigned an 8-hour TWA sound level based on noise measurements and job activities at the time of the survey. It is possible that larger errors in estimating 8-hour TWA sound levels over a long period of time may have occurred for workers with longer durations of exposure. It is also possible that the workers with long durations included in this study represented a population which may have been less sensitive to the adverse effects of noise on hearing. This may have contributed to the observed decrease in risk with increasing sound level, L_{NE}, for durations greater than 20 years. Hence, the cross-sectional design of the survey introduces areas of concern for predicting NIHH risks over a working lifetime.

B. Modeling caveats

The data limitations described above also placed limitations on the modeling approach and interpretations presented in this paper. One data limitation with implications for modeling the risk of noise-induced hearing handicap, was the lack of information on the distribution of 8-hour TWA sound levels among the control population. This is a crucial omission because all excess risk estimates depend on the risk of handicap among workers with low levels of occupational noise exposure (in this study, defined as exposure to sound levels less than 80 dB).

Due to this lack of data, a very simplistic assumption was made: sound levels in the control population could be represented by a single number. This is problematic in terms of model interpretation. First, it ignores the possibility that there may be a distribution of sound levels below 80 dB for this population. Second, this assumption results in a model that implies that the estimated value (L_0) is a threshold sound level at which no excess risk of noise-induced hearing handicap is predicted regardless of the duration of exposure. Hence, the statistical criteria used in model development are valid only if all of the controls were below a defined threshold.

These modeling issues underscore the fact that all models are likely to be dependent on assumptions used to account for uncertainty in the available data. This analysis did not model hearing threshold levels as a continuous variable. Therefore, calculation of NIPTS using these models are not possible. The analysis also did not extensively explore other possible shapes for the sound level function other than $(L_{NE}-L_0)^{\phi}$. Furthermore, modeling exposure duration as a categorical variable limits finer examination of the relationship of duration of exposure on risk of hearing handicap.

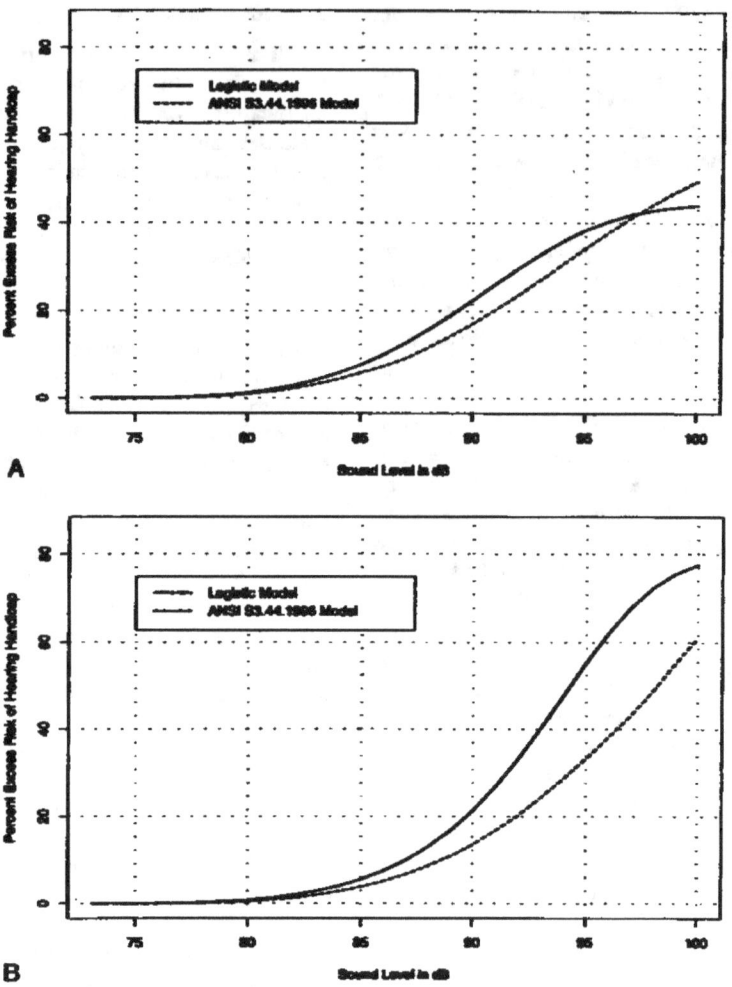

FIG. 5. Excess risk (A1-weighting, 1–4 kHz) as a function of sound level (L_{eq}) according to model 4 (Table IV), in comparison with curves derived from ANSI S3.44.1996 using Annex A database. Panel A: age 65 years, exposure duration 45 years. Panel B: age 45 years, exposure duration 25 years.

The models described in this paper were developed on the basis of this particular data set. Inferences based on the ONHS data set are also limited by its cross-sectional nature and the fact that exposure data was absent for the control population exposed to 8-hour TWA sound levels below 80 dB. As a result, the use of this model for other data sets with differing characteristics and different methods of data collection would not necessarily provide similar results.

C. Comparison of new risk estimates to ANSI S3.44

Given this updated analysis of the NIOSH (1972) data, it is of interest to compare these results to estimates generated using methodology developed by the International Standards Organization (ISO 1971, 1990), which was adopted in the ANSI S3.44 standard (ANSI, 1996). This standard was issued to provide a more accurate and more generalized model of the relationship between NIPTS and occupational noise exposure for people at different ages and duration of exposure. ANSI S3.44 (ANSI, 1996) provides mathematical procedures for estimating hearing handicap due to noise exposure for populations free from auditory impairment (other than that due to noise).

The data from studies by Passchier-Vermeer (1968) and by Burns and Robinson (1970) are the basis of the ANSI S3.44 (ANSI, 1996) standard for estimating NIPTS. As with the NIOSH (1972) study, most of the noise-exposed workers were exposed to daily noise levels ranging from 85 to 95 dB.

The Passchier-Vermeer (1968) and Robinson (1970) models are represented by different mathematical equations which include an aging (non-noise) component in dB and a NIPTS component in dB. For each model, the equation for NIPTS was determined by age correcting the noise-exposed workers' hearing threshold levels to get the NIPTS component. An empirical equation was developed for NIPTS in terms of noise level and exposure time. For each model, the aging and NIPTS components were combined to compute total hearing threshold level in dB (ANSI, 1996). A simple arithmetic average of the NIPTS values of Passchier-Vermeer and Robinson are used to predict NIPTS for ANSI

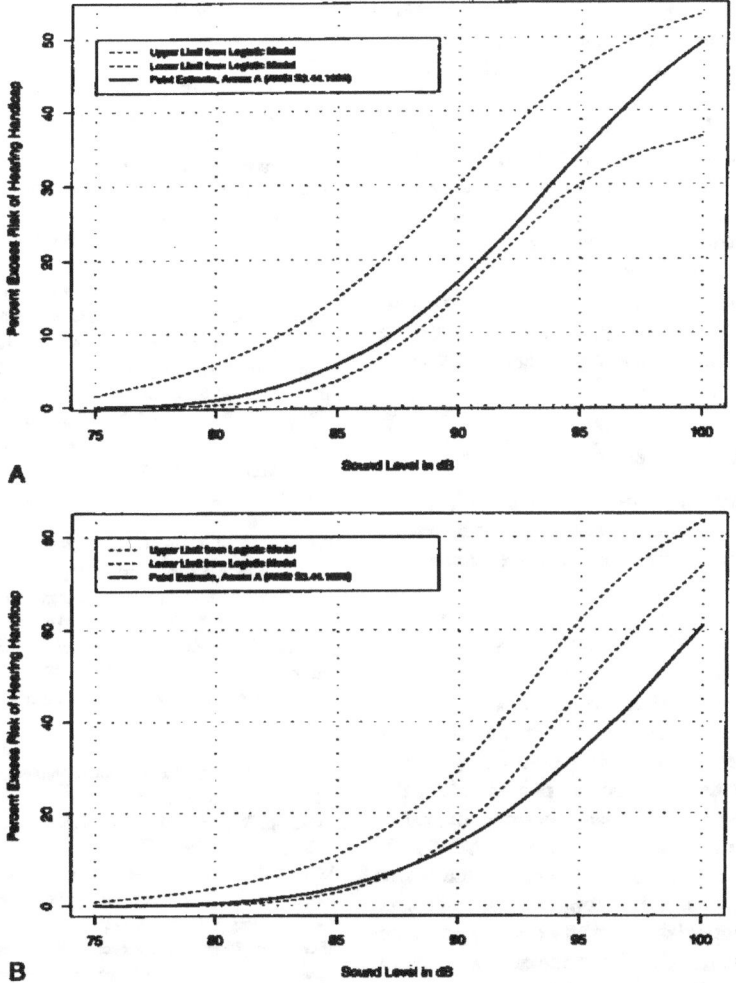

FIG. 6. Bootstrap-based 90% lower and upper confidence limits for excess risk (AI-weighting, 1–4 kHz) as a function of sound level (L_{AE}) according to model 4 (Table IV), in comparison with curves derived from ANSI S3.44.1996 using Annex A database. Panel A: age 65 years, exposure duration 45 years. Panel B: age 45 years, exposure duration 25 years.

S3.44 (ANSI, 1996). Johnson (1978) provides the methodology used to develop risk percent calculations using the percentage of the population expected to exceed a specific hearing threshold level (e.g., 25 dB) for a given population.

The excess risks generated from our analysis of the 1–4 kHz AI definition are compared to excess risk estimates generated using the ANSI S3.44 (ANSI, 1996) methodology and Annex A as the unexposed population. Annex A was chosen over Annex B since the NIOSH study population was highly screened. Hence, the Annex A highly screened control population is the most appropriate comparison to our study population. As shown in Fig. 5, excess risk estimates from our best fitting model are similar to those estimated by ANSI S3.44 (ANSI, 1996) for workers aged 65 years with 45 years of exposure. However, among workers aged 45 years with 25 years of exposure, excess risk estimates at sound levels greater than 90 dB are higher for this analysis as compared to ANSI S3.44 (ANSI, 1996). These results particularly in the range of 80–90 dB are not surprising given the similarities in study design, data collection and time period for all of these studies. Although these are qualitative comparisons, the differences in estimates of lifetime excess risk between ANSI S3.44 (ANSI, 1996) and this analysis do not appear to be substantial. This is illustrated in Fig. 6, which shows that excess risk estimates generated from ANSI S3.44 are located between the bootstrap-based 90% upper and lower confidence limits from the best fitting logistic model. At age 45 years and 25 years of exposure, excess risk estimates below 89 dB are within the lower bound of the confidence limits from the logistic model. Thereafter, point estimates from ANSI S3.44 are found to be lower, particularly at sound levels greater than 92 dB.

For other definitions of hearing handicap (0.5–2 kHz and 1–3 kHz), ANSI S3.44 estimates of excess risk are considerably lower at 85 dB for workers aged 65 years with 25 years of exposure. For the 0.5–2 kHz definition, excess risks at 85 dB from our logistic model and ANSI S3.44 (ANSI, 1996) are 12% and 1%, respectively. For the 1–3 kHz definition, the values are 16% for our model and 4% using ANSI S3.44 (ANSI, 1996) methods. At 80 dB, ANSI S3.44 generates excess risks of 0% for both definitions, while estimates from this analysis are 5% and 6% for the 0.5–2 kHz and 1–3 kHz definitions, respectively. Some of the divergent results may be due to differences in population characteristics of the studies used to generate excess risks. The NIOSH data set represented a heterogeneous population of workers from a variety of geographic regions and worksites within the United States. The study populations used to develop the ANSI S3.44 (ANSI, 1996) models were likely to be more homogeneous with respect to industry, demographic and socioeconomic (e.g., access to medical care) characteristics.

D. Future directions and data needs

This analysis indicates a need to collect and analyze data from populations exposed to noise at sound levels below 85 dB to learn more about the shape of the dose-response relationship below 85 dB. Like similar studies conducted in the late 1960 and early 1970's, the screened ONHS data set had few subjects with exposures at levels below 85 dB. This contributed to a high degree of instability in the risk estimates as the modeling assumptions were varied. Although logistic modeling techniques were used in this analysis, other methods for evaluating excess risks can reasonably be applied to these data. Nonetheless, it seems plausible that the observed instability below 85 dB would persist using other modeling methods. Risk estimates in the range of 88–95 dB are probably more reliable than the estimates for the lower ranges of sound level. More recent longitudinal data sets may be useful in examining risk below 85 dB. To examine whether noise-induced hearing handicap remains a problem for workers enrolled in OSHA-mandated hearing conservation programs (Department of Labor, 1981a,1981b), we are currently examining appropriate longitudinal audiometric databases. The present analysis indicates that new studies should be implemented to (1) characterize noise exposure for presumably "non-noise" or low noise populations (including populations exposed to nonoccupational sources of noise); and (2) examine dose-response relationships for noise and hearing handicap among workers exposed to noise levels below 90 dB.

ACKNOWLEDGMENTS

The authors wish to thank Dr. John Franks for his useful comments and advice in the development of this work and Barry Lempert for supplying an electronic version of the data for this analysis.

ANSI (1969). ANSI S3.5-1969, "American National Standard Methods for the Calculation of the Articulation Index (American National Standards Institute, New York).

ANSI (1996). ANSI S3.44-1996, American National Standard Determination of Occupational Noise Exposure and Estimation of Noise-Induced Hearing Impairment (American National Standards Institute, New York).

ASHA (1981). American Speech-Language-Hearing Association Task Force on the Definition of Hearing Handicap, "On the definition of hearing handicap," Asha 23, 293–297.

Baughn, W. L. (1973). "Relation between daily noise exposure and hearing loss as based on the evaluation of 6835 industrial noise exposure cases," AMRL-TR-73-53, Aerospace Medical Research Laboratory, Wright-Patterson Air Force Base, Ohio.

Breslow, N. E., and Day, N. E. (1980a). "Classical Methods of Analysis of Grouped Data," in *Statistical Methods in Cancer Research: Vol. 1– The Analysis of Case-control Studies* (International Agency for Research on Cancer, Lyon, France), IARC Publication No. 32, Chap. 4, pp. 140–148.

Breslow, N. E., and Day, N. E. (1980b). "Unconditional Logistic Regression for Large Strata," in *Statistical Methods in Cancer Research: Vol. 1–The Analysis of Case-control Studies* (International Agency for Research on Cancer, Lyon, France), IARC Publication No. 32, pp. 192–247.

Burns, W., and Robinson, D. W. (1970). *Hearing and Noise in Industry* (Her Majesty's Stationary Office, London).

Cohen, A. (1996). Personal communication. August 1996.

Department of Labor (1981a). 46 Fed. Reg. 11, "Occupational noise exposure; Hearing Conservation amendment, rule, and proposed rule, part III," pp. 4078–4179.

Department of Labor (1981b). "Final regulatory analysis of the hearing conservation amendment," Report number 723-860/752 1-3, U.S. Government Printing Office, Washington, D.C.

Efron, B. (1982). *The Jackknife, the Bootstrap and Other Resampling Plans* (Society for Industrial and Applied Mathematics, Philadelphia, PA).

Efron, B., and Tibshirani, R. (1986). "Bootstrap methods for standard errors, confidence intervals, and other measures of Statistical Accuracy," Stat. Sci. 1(1), 54–77.

Federal Register (1974). "Environmental Protection Agency comments on Proposed OSHA Occupational Noise Exposure Regulation," Vol. 39, No. 244, pp. 43802–43809.

Fienberg, S. E. (1987). *The Analysis of Cross-Classified Categorical Data* (MIT, Cambridge, MA).

ISO 1999 (1971). "Assessment of occupational noise exposure for hearing conservation purposes," First Edition, International Organization for Standardization, ISO/R 1999-1971.

ISO 1999 (1990). "Acoustics—Determination of occupational noise exposure and estimation of noise-induced hearing impairment," International Organization for Standardization.

Johnson, D. L. (1978). "Derivation of presbycusis and noise induced permanent threshold shift (NIPTS) to be used for the basis of a standard on the effects of noise on hearing," AMRL-TR-78-128, Aerospace Medical Research Laboratory, Wright-Patterson Air Force Base, Ohio.

Lempert, B. L., and Henderson, T. L. (1973). "Occupational Noise and Hearing 1968 to 1972: A NIOSH Study," U.S. Department of Health, Education, and Welfare, Public Health Service, Center for Disease Control, National Institute for Occupational Safety and Health, Division of Laboratories and Criteria Development, Cincinnati, OH.

NCHS (National Center for Health Statistics) (1965). "Hearing Levels of Adults by Age and Sex, United States, 1960–72," Vital and Health Statistics, Public Health Service Publication No. 1000-Series 11-Np. 11, U.S. Government Printing Office, Washington, D.C.

NIOSH (1972). "NIOSH criteria for a recommended standard: occupational exposure to noise," Cincinnati, OH: U.S. Department of Health, Education, and Welfare, Public Health Service, Center for Disease Control, National Institute for Occupational Safety and Health, DHSS(NIOSH) Publication No. HIM 73-11001.

Passchier-Vermeer, W. (1968). "Hearing loss due to exposure to steady-state broadband noise," Report No. 35 and Supplement to Report No. 35, Institute for Public Health Engineering, The Netherlands.

Robinson, D. W. (1970). "Relations between hearing loss and noise exposure," in *Hearing and Noise in Industry*, edited by W. Burns and D. W. Robinson (Her Majesty's Stationary Office, London), pp. 100–151.

Robinson, D. W., and Sutton, G. O. (1975). "A comparative analysis of data on the relation of pure-tone audiometric thresholds to age," NPL Acoustics Report AC84, England, April 1978.

Royster, L. H., and Thomas, W. G. (1979). "Age effect hearing levels for a white non-industrial noise exposed population (NINEP) and their use in evaluating hearing conservation programs," Am. Ind. Hyg. Assoc. J. **40**, 504–511.

SAS Institute, Inc. (1989). "SAS/STAT User's Guide," Version 6, Fourth Edition Volume 2. SAS Institute, Cary, NC, pp. 1071–1125, pp. 1135–1193.

Statistical Sciences, Inc. (1993). *S-PLUS for Windows Users Manual, Version 3.1* (Statistical Science Inc., Seattle).

Ward, W. D., and Glorig, A. (1975). "Protocol of inter-industry noise study," J. Occup. Med. **17**(12), 760–770.

Yerg, R. A., Sataloff, J., Glorig, A., and Menduke, H. (1978). "Inter-industry noise study," J. Occup. Med. **20**(5), 351–358.

www.ingramcontent.com/pod-product-compliance
Lightning Source LLC
Chambersburg PA
CBHW081728170526
45167CB00009B/3743